6th International Conference on Mechanical, Materials and Manufacturing (ICMMM 2019)

IOP Conference Series: Materials Science and Engineering
Volume 689

Boston, Massachusetts, USA
12 - 14 October 2019

ISBN: 978-1-7138-0981-4
ISSN: 1757-8981

Printed from e-media with permission by:

Curran Associates, Inc.
57 Morehouse Lane
Red Hook, NY 12571

Some format issues inherent in the e-media version may also appear in this print version.

Institute of Physics
Dirac House, Temple Back
Bristol BS1 6BE UK

Phone: 44 1 17 929 7481
Fax: 44 1 17 920 0979

techtracking@iop.org

Additional copies of this publication are available from:

Curran Associates, Inc.
57 Morehouse Lane
Red Hook, NY 12571 USA
Phone: 845-758-0400
Fax: 845-758-2633
Email: curran@proceedings.com
Web: www.proceedings.com

TABLE OF CONTENTS

ADVANCED MANUFACTURING SYSTEMS AND TECHNOLOGIES

IOP Conf. Series: Materials Science and Engineering **689** (2019) 011001 doi:10.1088/1757-899X/689/1/011001

Preface

As the Chair of 2019 6th International Conference on Mechanical, Materials and Manufacturing (ICMMM 2019), I would like to say it is my honor to have been involved with this excellent and exciting conference. Mechanical, materials and manufacturing are three disciplines that are all link directly. A conference that allows this latitude allows for research was success and the edge of knowledge can be pushed fully. This conference was no exception and I would like to thank the prestigious professors that supported the conference and presentations.

Many conferences claim to be international, very few are. In Boston, there were delegates for north and south America, Europe, Africa, middle east, Indian region, South Korea and China. This joining of worldwide researchers allowed for an excellent conference. Furthermore, posters were of a high quality in keeping with the selection process of the work.

Conferences need reviewers, and a thank you to all ours that supplied excellent feedback to maintain our standards. The level of presentation was as I expected, first class. These proceedings are to show the work and depth of research. We hope these will be the seminal work for larger research projects and cited as deserved. Finally, if you are reading and find interesting, submit a paper for next year, if you presented this year return and help us grow to deliver the support these disciplines deserved.

Warmest regards,

Professor Ian R. McAndrew
PhD FRAeS
ICMMM 2019 Conference Chair

IOP Conf. Series: Materials Science and Engineering **689** (2019) 011002 doi:10.1088/1757-899X/689/1/011002

Conference Committees

Conference General Chair

Ian McAndrew, Capitol Technology University, USA

Conference Local Organizing Committee

Haijun Gong, Georgia Southern University, USA

Conference Program Chairs

Young Moon, Syracuse University, USA

Nourredine Boubekri, University of North Texas, USA

Carl Moore, FAMU-FSU College of Engineering, USA

Conference Program Co-chair

Daniel Semere, KTH Royal Institute of Technology, Sweden

Conference Technical Committees

Liew Yun Ming, Unversiti Malaysia Perlis, Malaysia

Nur Farhana Diyana Mohd Yunos, Unversiti Malaysia Perlis, Malaysia

Dolores Eliche Quesada, University of Jaén, Spain

Norainiza Saud, Unversiti Malaysia Perlis, Malaysia

Vladimir Gurau, Georgia Southern University, USA

Hugo Miguel Silva, University of Minho, Portugal

Ching Yern Chee, University of Malaya, Malaysia

Linxia Gu, University of Nebraska-Lincoln, USA

Jan Górecki, Poznan University of Technology, Poland

Hong Jin Kim, Globalfoundries, USA

Chao Jia, Tianjin University of Technology, China

Giuseppe Vairo, University of Rome "Tor Vergata", Italy

Xiaodong Xing, Harbin Engineering University, China

Ramesh Chandra Gupta, Malaviya National Institute of Technology, India

Samir Mekid, King Fahd University Petroleum & Minerals, Saudi Arabia

Priyanka Tyagi, GD Goenka University, India

IOP Conf. Series: Materials Science and Engineering **689** (2019) 011002 doi:10.1088/1757-899X/689/1/011002

IOP Conf. Series: Materials Science and Engineering **689** (2019) 011002 doi:10.1088/1757-899X/689/1/011002

Doriana D'adonna, University of Naples Federico II, Italy

Habib Gürbüz, Süleyman Demirel University, Turkey

Josue Aaron Leyva, CETYS UNIVERSIDAD, Mexico

Peter Olubambi, University of Johannesburg, South Africa

IOP Conf. Series: Materials Science and Engineering **689** (2019) 011003 doi:10.1088/1757-899X/689/1/011003

Peer review statement

All papers published in this volume of *IOP Conference Series: Materials Science and Engineering* have been peer reviewed through processes administered by the proceedings Editors. Reviews were conducted by expert referees to the professional and scientific standards expected of a proceedings journal published by IOP Publishing.

IOP Conf. Series: Materials Science and Engineering **689** (2019) 012001 doi:10.1088/1757-899X/689/1/012001

Weak adhesion and delamination of hydrogenated diamond like carbon coating on a rough surface of tappet valve substrate

F O Kolawole[1,2*], L B Varela[1,3], S K Kolawole[4,5], M A Ramirez[6], A P Tschiptschin[1]

[1]Departamento de Engenharia Metalúrgica e de Materiais, Universidade de São Paulo, Brazil
[2]Department of Materials and Metallurgical Engineering, Federal University, Oye-Ekiti, Nigeria
[3]Vale Institute of Technology, Juscelino Kubitschek Av., 31, Bauxita, Ouro Preto, MG, Brazil
[4]National Agency for Science and Engineering Infrastructure, Abuja, Nigeria
[5]Materials Science and Engineering, African University of Science and Technology, Abuja, Nigeria
[6]Instituto de Pesquisa e Desenvolvimento, Universidade do Vale do Paraiba, São Jose dos Campos, Brazil
*Corresponding author: fkopresido@yahoo.com

Abstract. DLC(a-C:H) film are known for their excellent tribological properties however, possess poor adhesion on metallic substrate causing film delamination. Introduction of adhesive silicon interlayer has showed improvement in the adhesion of DLC-H on metallic substrate. In the present study DLC-H was deposited on tappet valve and silicon wafer using PECVD. The as-deposited DLC-H was characterized using XRD, EDS and Raman spectroscopy. SEM was also used to measure the thickness of the DLC-H coating and the average surface roughness (R_a) was measured using 3D Profilometry. Scratch test was used to determine the COF, adhesion and hardness of the DLC-H coating. The DLC-H coating on tappet valve had weak adhesion compared with DLC-H coating deposited on silicon substrate, which had good adhesion. The result of the Raman spectroscopy revealed a sp3 fraction of 0.43 for the DLC-H deposited on silicon substrate. The thickness of the as-deposited DLC-H film was 1.78 μm, which was measured using SEM to view the cross-section of the as-deposited DLC-H coating on the silicon substrate. The surface roughness of the as-deposited DLC-H coating on silicon substrate was 26.59 nm. Silicon substrate coated with DLC-H had the lowest COF, ranging from 0.06 to 0.09.

1. Introduction

In the past two decades DLC coatings has become very attractive due to their wide range of applications in various fields (mechanical, electrical, tribological, biomedical and optical fields) [1-2]. Friction and wear are the main cause of high energy consumption in automotive engine parts, making most processes inefficient. However, using DLC coatings helps in lowering friction and wear rate [3-4]. Tribological characteristics of DLC coatings can be affected by adhesion promoter intermediate layer, substrate surface roughness, hydrogen incorporation or hydrogen non-involvement and coating deposition parameters (bias voltage, etching, current, precursor gas, time and substrate temperature) [5]. DLC(a-C:H) film are known for their excellent tribological properties, but has low chemical affinity with metallic alloys and high intrinsic stress, leading to poor adhesion on substrate leading to film delamination. Introduction of adhesive silicon interlayer deposited at different temperatures and times by electrostatic confinement plasma enhanced chemical vapour deposition (EC-PECVD) has showed an improvement in adhesion on metallic substrate [6]. Results have showed that the adhesive interlayers produced from tetramethylsilane are chemical structured as (a-SiC_x:H). Also, the structure, chemical composition and thickness are all dependent on deposition conditions [6]. The thickness of the interlayers increases with deposition time and decreases with deposition temperature. The interlayer contains less hydrogen and silicon atoms at higher deposition temperature, with enhanced formation of Si-C bonds, which relates to the rise in the critical load values found for a-C:H film delamination when the a-SiC_x:H interlayers are deposited within 573K to 823K. While, the interlayers are containing less carbon atoms at higher deposition times, decreasing the critical load values of a-C:H film leads to delamination when the a-SiC_x:H interlayers are deposited from 5 to 10 minutes [6]. Weak adhesion and delamination from substrates, arising from high tangential stress and high internal compressive stress of a-C:H films, are main factors that limit the film thickness and their durability over time [7]. Hydrogen passivates the dangling bonds in the hydrogenated DLC films and permits only weak interactions between DLC and the sliding partner [8]. In this work, silicon interlayer followed by hydrogenated DLC was deposited on silicon wafer and rough surface of AISI 52100 steel substrate (tappet valve) using PECVD and characterized using XRD, SEM, Raman spectroscopy, 3D Profilometry and scratch test.

2. Experimental methods

2.1. Deposition of hydrogenated DLC on unpolished tappet valve

The tappet valve (AISI 52100) as seen in Figure 1, was first cleaned in an ultrasound water bath by placing the tappet valve into a beaker thereafter, liquid detergent and distil water were added and placed into the ultrasound water bath for a period of 10 minutes, then followed by alcohol for another 10 minutes. This was done in order to remove all contaminations like erase or particles that might affect the deposition of DLC-H on the tappet valve.

Figure 1. Image of tappet valve for GM Buick Chevrolet Daewoo.

Hydrogenated DLC coatings was deposited on an unpolished tappet valve and silicon wafer using Plasma Enhanced Chemical Vapour Deposition in an Ar/C_2H_2 atmosphere. The tappet valve and silicon wafer were placed in the PECVD reactor, argon was first used to clean using the following parameters: fixed Ar flow rate was 10 sccm, bias voltage applied to substrate 600 V, current 0.1 A for a duration of 10 minutes. Thereafter, silicone interlayer (SiH_4) was deposited to improve the adhesion (approx. 250 nm thick), two pure graphite targets were placed into the erosion zone. Silicone flow rate was 3 sccm, bias voltage applied to substrate 750 V, current 0.05 A for a duration of 10 minutes. DLC-H deposition flow rate was 3 sccm, bias voltage applied to substrate 700 V, current 0.05 A for a duration of 2 hours, hydrogen content (approx. 19% H_2). Standard tappet valve from a commercial GM Buick Chevrolet Daewoo (see Fig. 1). The tappet valve had an original roughness, Ra (5 μm), hardness, 4.2 GPa (~427.6 HV 0.05).

2.2. Coating characterization

XRD analyses was used for crystalline phases identification within the quantitative analysis using Rietveld. Philips - X'Pert PRO PW 3040/00 X-ray diffractometer with a Cu-anode, a generator voltage of 45 kV and a current of 40 mA with the goniometer 2θ values varying from 10 to 110° with an angular pitch of 0.02° at a scan rate of 100.0 s/step. High Score Plus software was used for the phase identification of the constituent of all the crystalline phases. The particle size was estimated using Debye Scherrer equation.

IOP Conf. Series: Materials Science and Engineering **689** (2019) 012001 doi:10.1088/1757-899X/689/1/012001

$$L = \frac{K\lambda}{\beta\cos\theta} \quad (1)$$

where L is particle size, λ is the X-ray wavelength in nanometer (nm), β is the peak width of the diffraction peak profile at half maximum height resulting from small crystallite size in radians, and K is a constant related to crystallite shape, normally taken as 0.9. The value of β in 2θ axis of diffraction profile must be in radians. The wavelength used for the samples was $\lambda Cuk\alpha 1 = 0.154178$ nm.

2.3. Scanning electron microscope (SEM)
The microstructure, morphology, thickness of the films, topography, backscattered electron images and EDS composition were obtained using Scanning electron microscope with field effect cannon - FEG - Inspect 50, with secondary electron detectors, backscatter, EDS X-ray spectrometer and EDAX backscattered electron diffraction chamber EBSD.

2.4. Raman spectroscopy
Raman spectroscopy, (Horiba scientific LabRAM HR Evolution system) having 633 nm excitation wave length laser source with a resolution of 1μm was used to analyse the DLC-H coating structure with respect to sp2/sp3 ratio. The extended and static modes were used to detect chemical compound formation and the carbon peaks (disorder, D and amorphous graphitic, G peaks) for the coating structure on the silicon substrate. The scan range for the identification of the carbon peaks were, 400 to 2400 cm^{-1}. Data was fitted with a Gaussian line shape in order to show the G and D peaks positions and the ratio of peak intensities. The ratio ID/IG was considered as an indicator of the carbon sp2/sp3 structure.

2.5. Surface profilometry
The surface roughness of silicon substrate and tappet valve both coated with DLC-H, were analysed using Profilometry 3D CCI, Taylor Hobson. Surface roughness data was analysed to the least square line, with Gaussian filter, 0.25 mm upper cut off and bandwidth.

2.6. Scratch test
Adhesion was analyzed for the coated tappet valve and silicon substrate through a critical load measurement, using a CETR UMT-2M-110 tribometer in linear reciprocated mode and the classic scratch test with a diamond tip, in which the applied load varied from 1 to 50 N, with a sliding speed of 0.1 mm/s for 90 secs. The load at which the coating was stripped from the substrate determined the critical load value. The hardness of the coating was also determined using;

$$HS = \frac{K \times F_N}{W^2} \quad (2)$$

Where K is a constant referring to the known geometry of scratching tip (Berkovic indenter k=2.31), F_N is applied normal (vertical) force, W is the width of scratch.

3. Results and discussion
The image in Figure 2a shows tappet valve coated with hydrogenated DLC, immediately after deposition while, Figure 2b shows the same tappet valve coated with hydrogenated DLC after 24 hours of deposition. It can be seen from Figure 2b that the hydrogenated DLC coating had weak adhesion to the tappet valve substrate, leading to delamination within 24 hours. The weak adhesion was as a result of the surface roughness of the tappet valve substrate. The surface roughness at the top of the tappet valve was different from the circular part of the tappet valve.

Figure 2. Image of tappet valve coated with hydrogenated DLC (a) immediately after deposition (b) 24 hours after deposition.

3.1. Characterization of hydrogenated DLC coating

The XRD diffractogram in Figure 3, shows result of the DLC-H coating on silicon substrate at 69.5°, a single peak was identified at 2theta, this indicates that it is a single crystal. The crystal size was calculated to be 23.04 nm, using Debye-Scherer equation (section 2.2). The EDS result (Table 1) revealed the elemental composition of the silicon substrate coated with DLC-H, tappet valve (upper, middle and lower sections) coated with DLC-H. The chemical composition for the tappet valve showed some variation depending on the part of the tappet valve having different surface roughness. The top, upper, middle and lower sections of the tappet valve had very little carbon content, which was as a result of delamination as seen in Figure 2.

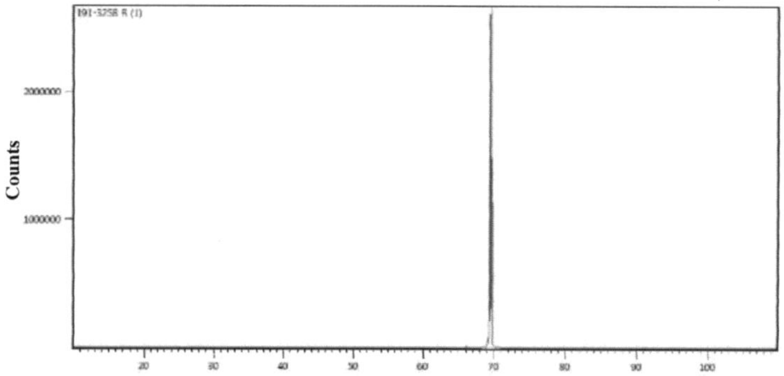

Figure 3. XRD diffractogram for silicon substrate coated with DLC-H.

3.2. Morphology and thickness of DLC-H coating

The upper section of the tappet valve shows very little DLC-H covering (circled red in Figure 4b). The middle section of the tappet valve had a better coverage of DLC-H, but shows poor adhesion to the tappet valve (Figure 4c). However, the lower section of the tappet, revealed that the DLC-H coating had good adhesion to the tappet valve (Figure 4d). The images in Figure 4b, c & d correlates to the carbon content as revealed from the EDS results (Table 1). Figure 5a, shows the SEM image of the top part of the tappet valve coated with DLC-H, it revealed that the top part had very low DLC-H content, thereby exposing the tappet valve (leading to high Iron (Fe) content) from the EDS result. While, Figure 5b, shows the SEM image for silicon substrate coated with DLC-H coating, which showed a good coverage due to strong adhesion of the DLC-H to the silicon substrate. Figure 5c, is the cross-section of the DLC-H coating, which has an average thickness of 1.79 µm as indicated in the SEM image, the light area of the image is the silicon substrate, while the dark area pointed by the red arrow is the thickness of the DLC-H coating.

Table 1. Chemical composition of various sections of DLC-H coated tappet and silicon substrate.

Substrate	C	Mn	Cr	Si	Ar	Fe
Silicon substrate	81.07	-	-	18.27	0.30	-
Top of tappet	2.30	0.52	1.02	7.47	-	85.43
Tappet upper section	-	0.62	1.10	1.20	-	97.08
Tappet middle section	4.38	0.42	1.21	5.96	-	88.04
Tappet lower section	82.23	-	-	2.06	0.32	11.72

IOP Conf. Series: Materials Science and Engineering **689** (2019) 012001 doi:10.1088/1757-899X/689/1/012001

Figure 6, shows the SEM image of the scratch track on tappet valve coated with DLC coatings. The image reveals the track due to the scratch on the DLC coating on silicon substrate by the diamond tip.

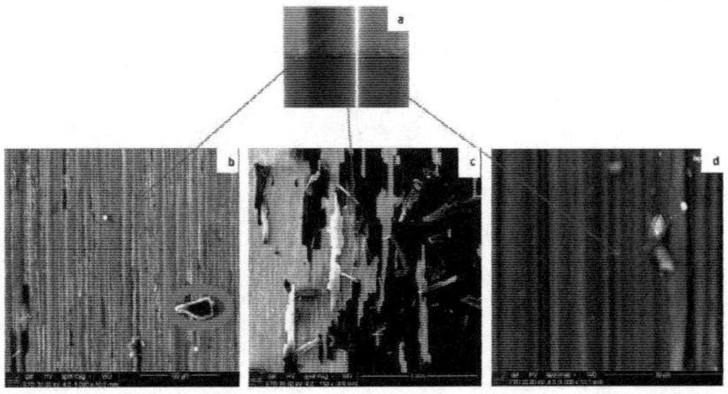

Figure 4. Tappet valve coated with DLC-H (a) image of coated tappet (b) SEM image of upper section (c) SEM image of middle section (d) SEM image of lower section.

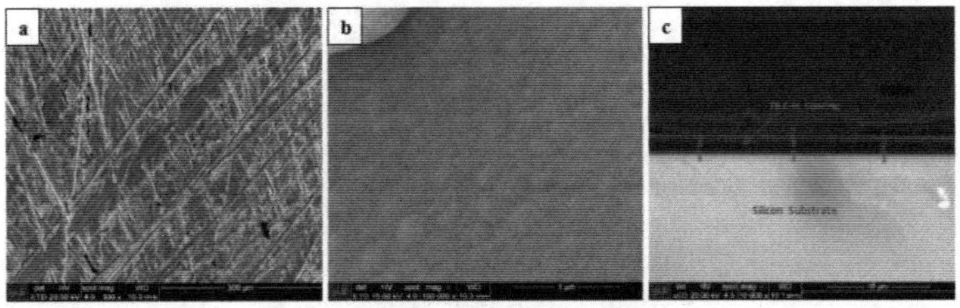

Figure 5. SEM image (a) top of tappet valve coated with DLC-H (b) silicon substrate coated with DLC-H (c) cross-section of DLC-H coating on silicon substrate.

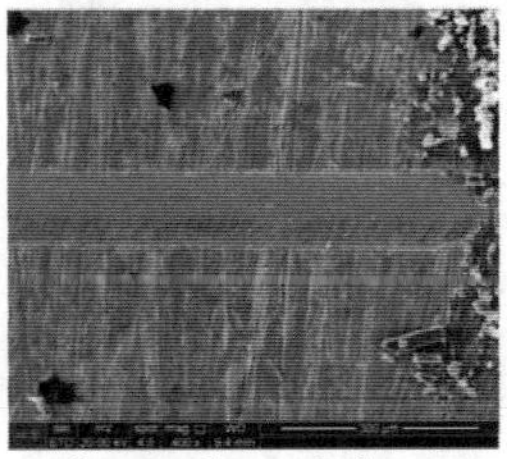

Figure 6. SEM image of scratch track for tappet valve with DLC coating.

3.3. Structure and Surface roughness of DLC-H

Raman which is the best and effective way to characterize the detailed bonding structure of DLCs. Graphite is a single crystal, having a Raman active mode at 1580 cm^{-1}, zone centred at E_{2g} symmetry labelled 'G' for graphite. Diamond also has a single Raman active mode at 1322 cm^{-1}, zone centred at T_{2g} symmetry. Graphite has a second mode at around 1350 cm^{-1} of A1g symmetry labelled 'D' for disorder, which corresponds to the breathing vibrations of rings at the K zone boundary (Robertson, 2002). The G mode is actually the stretching vibration of any pair of sp2 sites, for both carbon chains or aromatic rings. G does not only mean graphite, the D mode is the breathing mode of those sp2 sites only in rings, not in chains [9].

Also, the ratio of intensity of D/G peaks is a measure of the defects present on carbon nanomaterials structure. D/G ratio is related to sp3/sp2 carbon ratio, if carbon material is fully oxidized, i.e., all carbons are sp3. If D peak is higher, it means that the sp2 bonds are broken which in turn means that there are more sp3 bonds and more transition from sp2 to sp3 material, and there will be a maximum D/G ratio [10]. The Raman spectra plot in Figure 7 below has a Raman active mode at 1369 and 1570 cm^{-1} with intensity of 3970 and 5352 representing the D peak and G peak respectively. The Raman spectra for the DLC-H coating on silicon substrate has a I(D)/I(G) ratio of 0.74, comparing with the work of Ferrari and Robertson, 2000 [10], using the G position and I(D)/I(G) ratio vs sp3 fraction for as-deposited a-C:H we obtained a sp3 fraction of 0.43. The C-C network of a-C:H depends strongly on its hydrogen content, which links the sp2 and sp3 phases together. The average surface roughness (Ra) for the top of tappet valve and side of tappet valve and silicon substrate all coated with DLC-H were 405.90, 821.40, 26.59 nm respectively, the result shows that the silicon substrate coated DLC-H had the least surface roughness, which indicates proper adhesion to the silicon substrate.

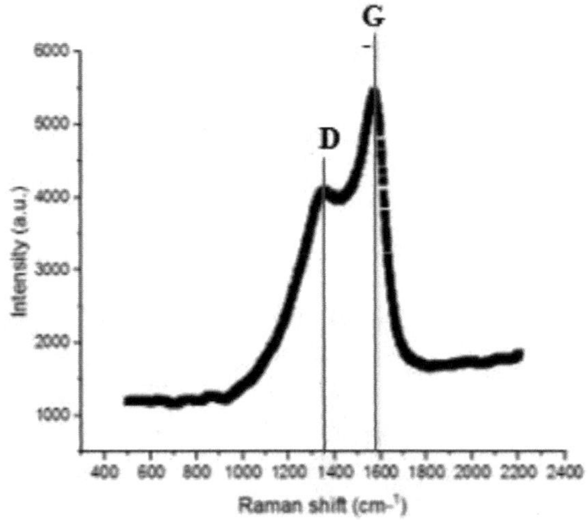

Figure 7. Raman spectra of the DLC-H coating on silicon substrate.

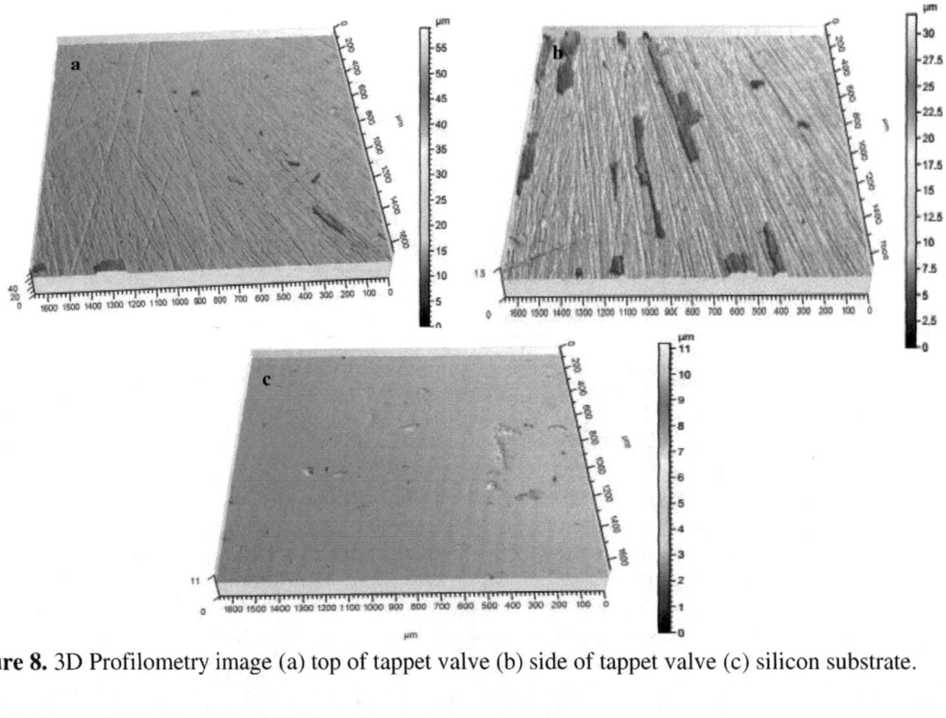

Figure 8. 3D Profilometry image (a) top of tappet valve (b) side of tappet valve (c) silicon substrate.

3.4 Adhesion and delamination of DLC coating

The coefficient of friction (COF) for tappet valve and silicon substrate coated with DLC can be seen in the plot (Figure 9). The COF for the tappet valve was measured at three different points having a range of 0.12 to 0.23, while that of the silicon substrate was 0.06 to 0.09, The silicon substrate fractured at

68 seconds, leading to an increase in COF. Small radial cracking of the DLC-H coating was observed on the surface, leading to tiny peaks at short intervals. The COF shows that the DLC-H coated silicon substrate had better adhesion than that of the tappet valve coated with DLC-H, this is as a result of the high roughness associated with the tappet valve substrate. Hardness value was determined to be 8.5 GPa, using equation 2 (section 2.6).

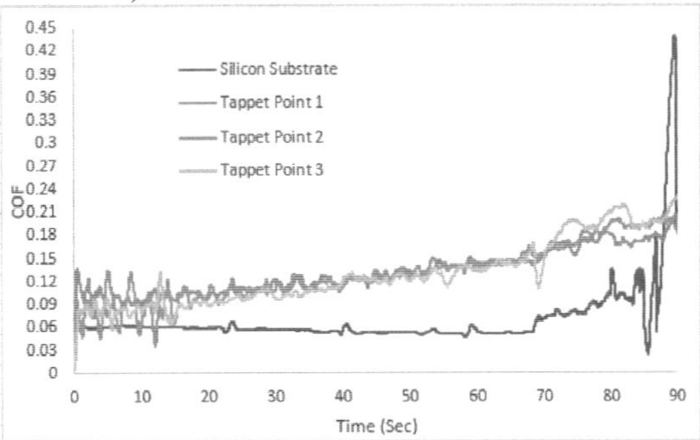

Figure 9. Coefficient of friction for tappet valve and silicon substrate coated with DLC.

4. Conclusion

In conclusion, hydrogenated DLC coating on tappet valve (AISI 52100) with high surface roughness possess weak adhesion, due to high compressive stress. However, for the DLC-H coating deposited on silicon substrate at the same conditions had a good adhesion. The result of the Raman spectroscopy revealed a sp3 fraction of 0.43 for the DLC-H deposited on silicon substrate. The thickness of the as-deposited DLC-H film was 1.78 μm, which was measured using SEM to view the cross-section of the as-deposited DLC-H coating on the silicon substrate. The surface roughness of the as-deposited DLC-H coating on silicon substrate was 26.59 nm. Silicon substrate coated with DLC-H had the lowest COF, ranging from 0.06 to 0.09 while, the tappet valve coated with DLC-H ranging from 0.12 to 0.23. This research work shows that the silicone interlayer improved the DLC-H adhesion on silicon substrate, however, it was not the same case for the tappet valve, due to high surface roughness.

Acknowledgement
The authors would like to thank, Petroleum Technology Development Fund (PTDF) and CNPq for its financial support. Also, the authors are grateful to LFS, Polytechnic School of Sao Paulo University for the use of Raman Spectroscopy and 3D Profilometry.

References

[1] Ladwig A M, Koch R D, Wenski E G, Hicks R F 2009 *Diamond & Rel. Mat.* **18** 1129.
[2] Suzuki M, Saito T, Tanaka A 2013 *Wear*, **304** 83.
[3] Mobarak H M, Chowdhury M 2014 Tribol. in Indus. **36** 163.
[4] Erdemir A, Holmberg K 2015 *Coating technology for vehicle applications*. Cham, Switzerland, Springer International Publishing AG 10-11.
[5] Al Mahmud K A H, Kalam M A, Masjuki H, Mobarak H M, Zulkifli N W M 2014 *J. rev. in solid state & mat. sci.* **40** 90.
[6] Cemin F, Bim L T, Menezes C M, Maia da Costa M E H, Baumvol I J R, Alvarez F, Figueroa C A 2015 *Surf. & Coat. Techn.* **283** 115.
[7] Sheeja D, Tay B K, Leong K W, Lee C H 2016 *Diamond & Rel. Mat.* **11** 1643.
[8] Grill A 1993 *Wear* **168** 143.
[9] Robertson J 2002 Mat. Sci. & Eng. **R37** 129.
[10] Ferrari A C, Robertson J 2000 Phys. Rev. B, **61** 14095.

IOP Conf. Series: Materials Science and Engineering **689** (2019) 012002 doi:10.1088/1757-899X/689/1/012002

Mechanical properties of self-compacting concrete containing sandstone slurry

R C Gupta, P Basu and S Agrawal
Department of Civil Engineering, Malaviya National Institute of Technology, Jaipur-302017, India

Email: rcgupta.ce@mnit.ac.in; sprarthitas@gmail.com; 2016uce1174@mnit.ac.in

Abstract. India embraces wide variety of dimensional stones. It includes granite, marble, sandstone, limestone and slate. Amid all the states, Rajasthan is known as mineral majestic of India, with an abundance of important stones. Rajasthan produces more than 90% of sandstone of all over the country. The total quantity of mined out sandstone is 200.27 million tons per year, out of which 110.25 million tons are accounted as waste during various manufacturing and processing stages of sandstone. An experimental work has been conducted to manufacture self-compacting concrete (SCC) with dry sandstone slurry (DSS) and study the properties of SCC. Seven SCC mixes have been prepared with various percentages of sandstone slurry (i.e. 0, 5, 10, 15, 20, 25 and 30 percent) as partial replacement of pozzolana portland cement (PPC). Compressive, flexure and tensile strength of all SCC mixes up to 180 days curing age and abrasion resistance after 28 days curing age have been found out and presented in this paper. The conclusion of this study draws an idea about the use of SCC mixes made with DSS are in favor of using DSS up to 10% for reinforced cement concrete work and up to 20% for plain cement concrete work.

1. Introduction

India is the largest producer of natural stone in the world. This country accounts for about 27 percent of the world's total natural stone production [1]. Rajasthan is situated in the western region of the country with an area of 3,42,238 km^2, which makes this state as the largest state and second most mineral-rich state in India. It has a wide range of mineral deposits containing 64 different types of major and minor minerals and producing an annual revenue of more than 6000 million Indian Rupees [INR] [2]. There are 3403 mining leases for major minerals, 11861 mining leases for minor minerals and 18249 quarry licenses in the state [3]. The quantity of stone export has been observed significant growth and it rises to almost 50 percent in a year 2011–12 [4]. The rapid flow in exports of sandstone in the last two decades has led to increasing numbers of sandstone quarrying and processing units. There are several stages through which sandstone wastes are generated. From table 1, it is visible that after mining, almost 50% of the dimensional sandstone ends up as waste. This extreme mining of dimensional stone has been originated a severe environmental influence. After drying, these slurry results in air pollution. Fine particles get entrained into the atmosphere due to turbulence from the highway traffic. Which not only reduces visibility but also creates respiratory problems. It has excellent nuisance value as it gets airborne even with a mild breeze and settles down in areas far away from the source. During the execution process of sandstone, numerous sandstone quarry workers expire from silicosis, a deadly but avoidable lung disease. It is due to the inhalation of dust containing crystalline silica during mining and processing [5]. Cutting and polishing waste discarded into nearby roads and dry rivers. In the rainy season, it makes the road surface slippery, which is dangerous for traffic. Disposal by dumping into rivers or sea is not safe, as it creates silting and a host of other problems for human and aquatic life. Therefore, it is essential to find an alternative use of DSS to safeguard the environment.

Table 1. Output of Mined-Out Sandstone Reserves in Rajasthan, India[6].

Sandstone waste generated [thousands of tons]	
Mined out reserves	200,272
Mine waste @ 25% of mine production	50,068
Sandstone production as per DMG Rajasthan[a]	150,205
Dressing waste	22,531
Processing + Polishing waste	37,651
Dressing waste + Polishing	60,182
Total waste	110,250

[a] Department of mines and geology (DMG) Rajasthan

2. Materials and experiments

2.1. Raw materials

In this investigational work, PPC has been used as a prime binder and partially replaced with DSS. The PPC used is as per IS 1489:1991 (part 1) [7] and DSS has been used after sieving through 90-micron sieve. Physical properties of PPC & DSS are shown in table 2. The particle size distributions for PPC & DSS (as obtained from mastersizer-3000) are presented in figure 1. Nearby river sand and crushed angular basalt rock (10mm maximum nominal size) have been used as fine and coarse aggregate respectively, and their physical properties are indicated in table 3. Sieve analysis has been performed as per IS 2386:1963 (part 1) [8] and its gradation has been designated as per IS 383:1970 [9]. Polycarboxylic ether (PCE) based superplasticizer (SP) as per IS 9103:1999 [10] and chloride free viscosity modifying agent (VMA) as per European Guidelines for Self-Compacting Concrete: Specification (EFNARC) Guidelines [11] have been used.

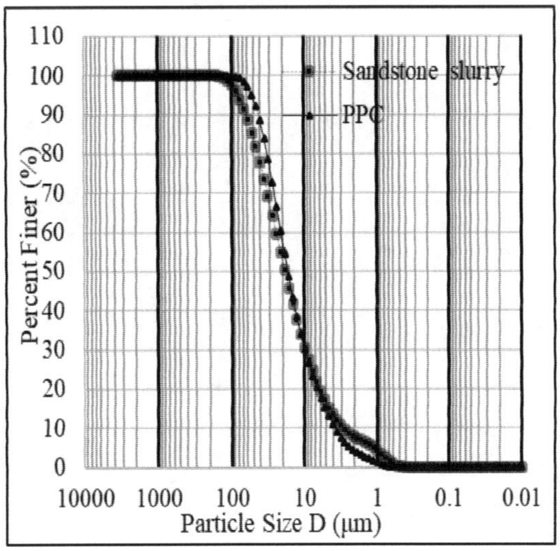

Figure 1. Particle size distribution curve of DSS and PPC

2.2. Mix proportions

Seven SCC mixes have been made in different proportion of PPC and DSS, by maintaining water/binder (w/b) ratio as 0.33 for 0, 5, 10, 15 and 20 percent replacement and 0.35 for 25 and 30 per cent replacement level. Quantity of fine aggregate, coarse aggregate and VMA are 807, 776 & 0.60 kg/m^3 respectively. The detail mix proportions, along with fresh property test results, are presented in table 4.

Table 2. Physical property of PPC and DSS.

Physical property of PPC		Physical property of DSS	
Description	Test results	Description	Test results
Colour	Grey	Colour	Pink
Specific gravity [g/cm^3]	2.91	Specific gravity, [g/cm^3]	2.62
Specific surface area [m^2/kg]	730.80	Specific surface area [m^2/kg]	968.70
Initial setting time [Minutes]	138.00	Liquid limit (WL) [%]	28.20
Final setting time [Minutes]	251.00	Plastic limit (WP) [%]	19.76
3-day comp. strength, [MPa]	24.41	Shrinkage limit [%]	27.11
7-day comp. strength, [MPa]	28.08	Shrinkage ratio	1.51
28-day comp. strength, [MPa]	37.30	Volumetric shrinkage	1.24

Table 3. Physical properties of coarse and fine aggregates.

Physical property	Coarse aggregate	Fine aggregate
Specific gravity [g/cm^3]	2.74	2.63
Water absorption [%]	0.18	0.50
Fineness modulus	5.94	2.61
Compacted bulk density [g/cc]	1.55	1.82
Percentage voids after Compaction	43.74	30.93
Loose bulk density [g/cc]	1.36	1.70
Percentage voids at the loose condition	50.42	35.36

Table 4. Mix proportions and fresh property of self-compacting concrete.

Mix	Quantity (kg/m^3)				Fresh property	
	Water	Powder		SP	Slump flow (mm)	T_{500} (sec)
		PPC	DSS			
DSS0	198	600	0	8.10	826.56	1.40
DSS5	198	570	30	8.40	818.92	1.43
DSS10	198	540	60	8.70	810.85	1.62
DSS15	198	510	90	9.00	800.71	1.79
DSS20	198	480	120	9.30	784.28	1.88
DSS25	210	450	150	9.60	781.40	1.91
DSS30	210	420	180	9.90	766.87	2.23

2.3. Laboratory testing program

2.3.1. Slump-flow and T_{500} test. The sampling and testing of fresh properties of self-compacting concrete have been conducted by following EN 12350-1 [12] and EN 12350-2 [13]. Test procedure is as follows: The fresh concrete is transferred into a cone which is used for slump test. When the top-level of slump cone is filled with concrete it is withdrawn upwards and the time from slump cone lifted upwards to when it flows and reaches a diameter of 500 mm is noted and is the T_{500} time. The maximum diameter of the slump flow is then measured with right angles, and the average reading is pointed out as the slump-flow reading [14].

2.3.2. Compressive strength test. This test has been conducted as per IS 516:1959 [15] on concrete cubes of 100 mm size after 7d, 28d, 90d and 180days of curing. Gradual increasing load at the rate of 140 kg/cm^2/minute is applied till the specimen fails. The maximum load is noted at which the specimen failed. The compressive strength is considered by dividing the load at the failure by the cross-sectional area of the sample.

2.3.3. Flexural strength test. This test has been conducted as per IS 516:1959 [15] on beam size 100 mm x 100 mm x 500 mm after designated curing time. Two-point loading set up is used, and the load is applied at the location of L/3 distance apart. The loading rate has been kept at 180 kg/minute. The load is applied gradually until the specimen reaches its failure load, and the load is recorded. Flexure strength is known as modulus of rupture.

2.3.4. Split tensile strength test. This test has been conducted as per DIN 1048:1991(Part 5) [16] on concrete cube specimen of 150 mm size. After specified curing days, cubes are taken out, and surface water has been wiped off with a dry cloth, for further testing process.

2.3.5. Abrasion resistance test. The abrasion test has been performed after 28-day of curing as per Indian Standard IS 1237:1980 [17]. Three 100 mm concrete cube specimens of each mix are oven dried at temperature 110 ± 5 °C for 24 h. After specimens cool down to room temperature, initial weight (W1) has been measured and after completion of the test, final weight (W2) has been taken. The change in thickness is calculated with the use of formula [Eq. 1] mentioned below:

$$t = \frac{W1-W2}{W1\ X\ A}\ X\ V1. \tag{1}$$

Where, V1 is the initial volume of the specimen in mm^3 and A is the surface area of the specimen in mm^2.

3. Result and discussions

3.1. Slump-flow and T_{500} test

The targeted slump flow was 800± 50 mm, which is classified to slump flow class 3 (SF3) (760 – 850 mm) as per EFNARC (2005). From slump flow result figure 2 it is noticed that the maximum slump flow diameter has been achieved at 0% replacement level and minimum slump flow diameter has been at 30% replacement level. In the present study, PPC has been used, which contains fly ash in it. Sandstone slurry particles are angular in shape. The increment of DSS increases the number of angular particles, which resist the movement of concrete flow. The almost similar trend of result has been reported by V. Corinaldesi [18], where Fly Ash (FA), Limestone powder (LP) and Rubber powder (RP) was used as mineral addition. The result showed that, increases in flow diameter of SCC due to the increment of FA having rounded particles. Whereas, LP and RP were in angular in size, which restricts the flowability of concrete.

3.2. Compressive strength test

The compressive strength test results have been based on 7d, 28d, 90d and 180 days strength test and results are presented in figure 3. At 7d, as the quartz sandstone slurry replacement gets increased, the substantial decrease in compressive strength has been noticed. This could be attributed to delay in cement hydration at the initial stage of curing due to the higher replacement level of quartz sandstone

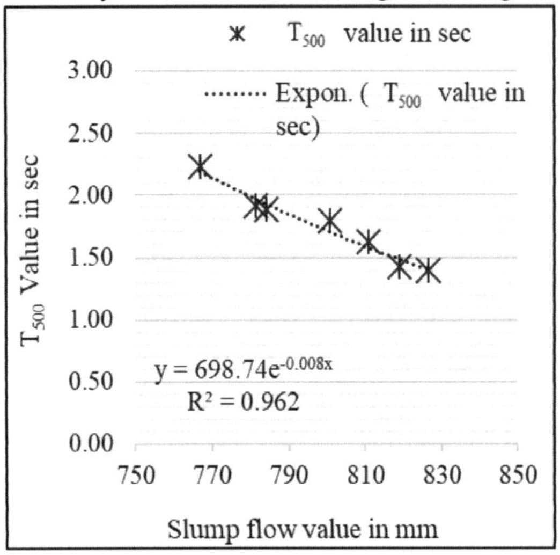

Figure 2. Slump flow v/s T_{500} result

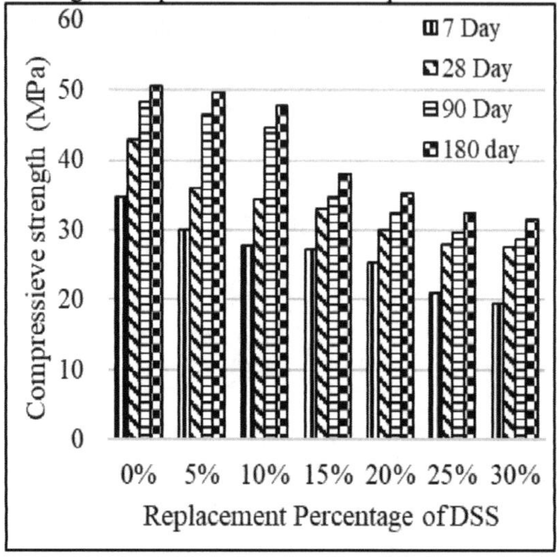

Figure 3. Compressive strength at 7d, 28d, 90d & 180days

slurry. After 90 days, significant strength gain has been observed and this is due to the continued pozzolanic reaction. This strength gain may also have been associated with filling up of pore spaces in the concrete with the product of hydration. As per Uysal et al. [19], larger specific surface area and siliceous property of waste material may be acted as nucleation sites for the early reaction products. As, PPC cement was used, which is having class F, fly ash and addition of DSS, which is having more than 90% of SiO_2, increases the silica content in SCC. The SiO_2 consumes the early formed Ca $(OH)_2$. Which is parting lesser amount in future ages for fly ash and it is causing early compressive strength. Therefore, it provides better strength at early stages to the cement due to the formation of silicates. This nucleation, accelerating the hydration process and significantly increase the C_3S content. As a result, the increase in compressive strength of the concrete has been observed in the early stage.

3.3. Flexure strength test

The flexure strength test performed at 28d, 90d and 180 days and results are shown in figure 4.

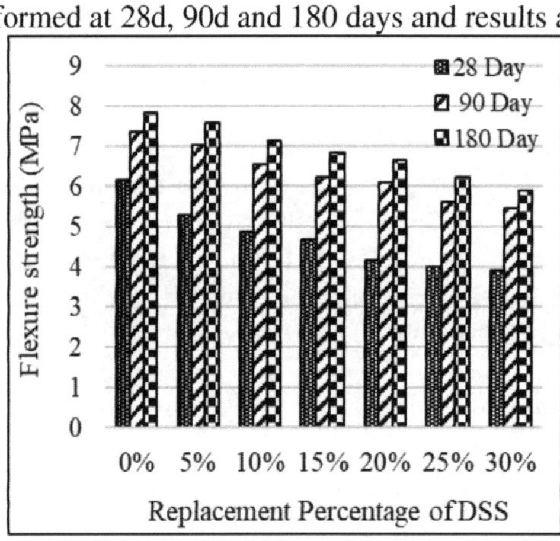

Figure 4. Flexure strength at 28d, 90d & 180days

At 0% replacement of DSS, flexure strength has been observed maximum with cement content 600 kg/m^3. At 30% replacement level of DSS lowest strength has been achieved where cement content is 420 kg/m^3. The almost the same pattern has been reported in Dina M. Sadek [20]. They reported that strength reduction in control self-compacting concrete is related to the less cement content, poor workability and consequently weak compactness of SCC.

3.4. Tensile strength test
The tensile strength test results have been determined at 28d, 90d and 180 days and presented in figure 5. Tensile strength shows good result for all the replacement level of DSS. This is in line with Dina M. Sadek report [20]. Wherein, they stated that using of waste stone powders as mineral additives in SCC enhances its physical, mechanical properties and durability-related properties compared with the control mix of the same cement content.

3.5. Abrasion resistance test
The wear and tear of concrete occurs due to abrasion under different loading conditions. The results obtained for different replacement level of DSS in SCC are shown in figure 6. With the incorporation of 30% sandstone slurry, depth of wear increases up to 1.93 mm, which shows an increment of 0.95 mm with respect to control SCC mix sample. As per IS 1237:1980, permissible wear of the individual specimen for using in general purpose tiles is 4 mm and for heavy duty floor tiles is up to 2.5 mm. For all the SCC mix depth of abrasion is within the permissible limit. However, this developed SCC has good abrasion resistance.

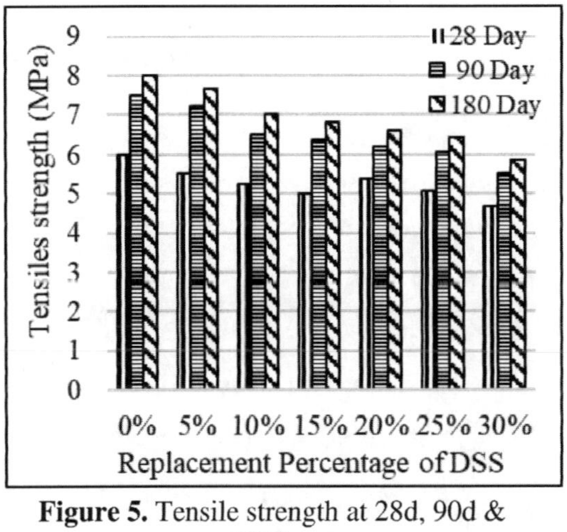

Figure 5. Tensile strength at 28d, 90d & 180day

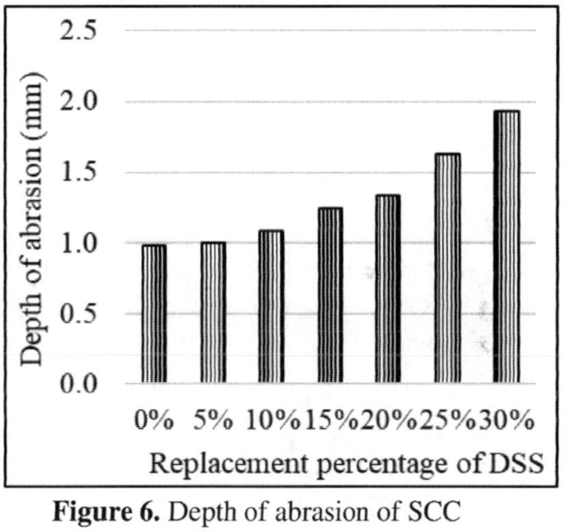

Figure 6. Depth of abrasion of SCC

4. Conclusions
The following conclusions can be drawn based on the above study: 1) All SCC mixes are classified in SF-3 category. This concrete may be beneficial for vertical applications in very congested structures and structures with complex shapes. 2) Due to the angular shape of sandstone slurry viscosity of SCC was increased and all the SCC mixes are classified into VS1 class, except DSS30 which contains 30% of dry sandstone slurry. 3) From strength parameter, it is visible that up to 10 per cent replacement level of DSS compressive strength, flexure strength and tensile strength parameters show excellent results. It happens due to enough quantity of binder content is present in the mixes. The sandstone slurry particle size is less than cement so better packing in between concrete constituents may be one of the reasons for achieving good strength. This mix can be used for reinforced concrete work. 4) Further increment of sandstone slurry up to 20%, concrete strength decreased, although it could be useful for plain cement concrete applications. 5) Depth of wear of concrete was increased with the increment of dry sandstone slurry percentages. However, for all the SCC mixes depth of wear lies within the permissible limit of standard code.

References

[1] Madhavan P and Dr. Raj S *Budhpura ground zero sandstone quarrying in India* Study commissioned by India Committee of the Netherlands, Mariaplaats 4e, 3511 LH Utrecht The Netherlands (2005) p.1-32

[2] Information on, http://www.dmg-raj.org/why-rajasthan.html

[3] Information on, http://resurgent.rajasthan.gov.in/focus-sectors/minerals-ceramics

[4] Crimes S B, Marshall S and Taylor K *Rajasthan stone quarries promoting human rights due diligence and access to redress in complex supply chains* (2016) p.1–67

[5] Lahiri-Dutt K *Digging to survive women's livelihoods in South Asia's small mines and quarries* South Asian Surv. **15** (2008) p.217– 44

[6] Kumar S, Gupta R C, Shrivastava S and Csetenyi L J *J. Mater. Civ. Eng.* **29** (2017)

[7] IS 1489 (part-1): 2015 *Portland pozzolana cement-specification Part-1 fly ash-based* Bureau of Indian Standards New Delhi India (2015)

[8] IS 2386 (Part I):1963 *Method of Test for aggregate for concrete Part I - Particle size and shape* Bureau of Indian Standards New Delhi India (1963) (Reaffirmed 2002)

[9] IS 383: 1970 *Specification for coarse and fine aggregates from natural sources for concrete* Bureau of Indian Standards New Delhi India (1970)

[10] IS 9103: 1999 *Specification for concrete admixtures* Bureau of Indian Standards New Delhi India (1999)

[11] EFNARC, EFCA *Guidelines for viscosity modifying admixtures for concrete* Viscosity Modifying Admixtures Concr (2006)

[12] BS EN 12350-1:2009 *Testing fresh concrete - Part 1: Sampling* (2009)

[13] BS EN 12350-2:2000 *Testing fresh concrete Part 2: Slump* (2000)

[14] The European Project Group *the European guidelines for self-compacting concrete: specification, production and use* Eur. Guidel. Self-Compact. Concr. (2005) 63

[15] IS 516: 2018 *Method of tests for strength of concrete* Bureau of Indian Standards New Delhi India (2018)

[16] DIN 1048: 2016 *Determination of permeability of concrete* German Standard (2016)

[17] BIS 1237: 2012 *Cement concrete flooring tiles specification* Bureau of Indian Standards New Delhi India (2012)

[18] Corinaldesi V and Moriconi G *Constr. Build. Mater.* **25** (2011) 3181–86

[19] Uysal M, Yilmaz K and Ipek M *Constr. Build. Mater.* **27** (2012) 263–70

[20] Sadek D M, El-Attar M M and Ali H A *J. Clean. Prod.* **121** (2016) 19–32

Rheological properties of two stainless steel 316L powders for additive manufacturing

Haijun Gong[1], Xiaodong Xing[2], Hengfeng Gu[3]

[1]Manufacturing Engineering Department, Georgia Southern University, Statesboro, GA 30458, USA
[2]College of Mechanical and Electrical Engineering, Harbin Engineering University, Harbin 150001, China
[3]ANSYS, Inc, Park City, UT 84098, USA

Email: hgong@georgiasouthern.edu; xingxiaodong@hrbeu.edu.cn; hengfeng.gu@ansys.com

Abstract. This study measures the rheological properties of two stainless steel 316L powders which are used for the powder-bed-fusion based additive manufacturing process. The purpose is to evaluate the newly acquired powder in comparison with the used and recycled powder, so that both powders can be mixed with each other to supplement the powder usage. The powder rheology properties, such as dynamic property, bulk property, and shear property, are tested and compared. The results and analysis confirm the compatibility of powder mixing.

1. Introduction

The Additive Manufacturing (AM) is a process of joining materials to make objects from 3D CAD model, usually layer upon layer [1]. This technology has been widely used for metal part production. The commonly used metal AM process consists of selective laser melting (SLM), laser cusing, laser engineered net shaping (LENS), electron beam melting (EBM), etc. The pre-alloy powders, such as stainless steel, cobalt chromium, titanium alloy, etc., can be directly used by these metal AM processes to produce functional parts, in contrast to the conventional molding or powder metallurgy process. In particular, the SLM and EBM processes, which are two popular powder-bed-fusion (PBF) based AM processes, have been extensively employed for the metal part fabrication. The characteristics of powder for PBF are the key for the process performance and part quality [2]. Many studies have pointed out the importance of understanding the powder effects on the PBF process, with the purpose of ensuring the production stability using AM. Otherwise, a profound understanding to the powder will help identify the powder quality so that the users can determine whether or not to keep using or discard the recycled powder, in order to maximize the material usage.

Some standardized testing methods have been applied to the PBF powders, such as the measurements of tap density, particle size distribution (PSD), flow through an orifice, powder morphology, and angle of repose (AOR). The testing results provide some insights about powder performance in the PBF process in correlation with the powder characteristics. Gu et al. [3] measured the AOR and PSD of 3 types of Ti-6Al-4V powders from different suppliers and concluded that the flowability is significantly influenced by the fine particle inclusions. Also, the thermal conductivity of powder bed is higher because of the fine particles. Tang et al. [4] studied the PSD, apparent density, tap density, flowability, and particle morphology of reused Ti-6Al-4V powder of EBM process. It is found that the reused powder exhibits improved flowability and modified PSD and particle morphology. Spierings et al. [5, 6] tested the properties of parts produced using stainless steel 316L powders with three different PSD and pointed out PSD should be optimized for better results.

Slotwinski et al. [7] characterized the size and shape of virgin powder and recycled powder, using X-ray computed tomography, optical microscopy, and scanning electron microscopy. The particle size and shape have a critical impact on forming agglomerates and irregular porous surface of metal AM parts [8]. Spierings et al. [9] also reviewed different powder flow measurement techniques for PBF based AM processes. However, researchers also realized that it is challenging to interconnect the properties of the individual powder particle with bulk powder behavior, in-process performance and finally part property because some powder particle properties can have contrary effects on the next-level properties [2].

Recently, powder rheometer has been of interest to be employed for the powder characterization. Strondl et al. [10] used a powder rheometer to determine various powder flow and powder bed properties. The study concluded that powder rheology is a useful method to determine differences in flow properties of powders used in AM. Lyckfeldt et al. [11] also demonstrated the versatility of powder rheometer on characterizing the properties of metal powders and correlated the powder cohesion, compressibility, aeration effect, and the yield stress at shear. Clayton et al. [12] use some case studies to illustrate how powder rheology supports optimization and lifecycle management of powders for AM. These studies exhibit promising potentials of utilizing powder rheology measurement to characterize metal powders for PBF process. Hence, in this study, the rheological properties of two stainless steel 316L powders are measured and compared. One powder is used and recycled for a couple of times. The other powder (virgin powder) is newly acquired for supplement the deficiency of the first powder. In order to identify the difference and achieve a repeatable performance of powders for metal AM, rheological parameters of both powders, such as stability index, specific energy, aeration ratio, etc., are measured and discussed.

2. Materials and Experiment

Two stainless steel powders were measured for this study. The used and recycled powder FS 316L came with the Farsoon FS271M SLM additive manufacturing machine which was procured from Farsoon High Tech Co. Ltd. The virgin powder MetcoAddTM 316L-A was ordered from Oerlikon AM, which is a metal powder supplier for AM. The chemical compositions and PSD of the two powders were provided by the powder supplier, as shown in Table 1 and 2. Both powders were produced by the gas atomized method and have a spherical morphology, with the apparent density larger than 4 g/cm^3.

Table 1. Chemical composition of stainless steel 316L powders.

	Weight Percent (wt. %)						
	Fe	Cr	Ni	Mo	Mn	C	Other
FS 316L	Balance	18.0	12.0	2.0	< 1.0	<0.03	<1.0
MetcoAdd 316L	Balance	17.4	11.2	2.6	1.5	<0.02	<1.0

Table 2. Particle size distribution of stainless steel 316L powders.

	D10*	D50	D90
FS 316L	22	34	52
MetcoAdd 316L	19	30	46

*Units: μm

The powder rheology measurements, i.e. dynamic property, bulk property, and shear property were conducted using a *Freeman* FT4 Powder Rheometer. As for the dynamic property, the rheometer uses a precision blade (impeller) to move downwards, with a certain revolving speed, through the powder to establish a precise flow pattern. The powder particles start to interact, or flow relative to one another. The resistance experienced by the blade represents the difficulty of this relative particle movement, or the bulk flow properties. The vertical force on the base and the torque on the blade are measured. The bulk property is determined by applying a normal force to the powder bed using a vented piston which allows entrained air to escape. Shear property provides an indication of how easily a powder will move from a static condition into dynamic flow. The test applies a known normal stress to the powder, followed by a rotational torque to determine the point of failure.

3. Results and Discussion

The chemical composition of both powders shows a similar elemental inclusion, which conforms to the standardization of the stainless steel 316L alloy. The PSD indicates that the average size of FS 316L powder is a little larger than the MetcoAdd 316L powder. The powder with a small particle size might present a decreased flowability and a higher specific surface area, causing agglomeration [13]. This is the major concern of mixing both powders to supplement the powders usage for AM. The following sections compare and discuss the rheological measurement of both powders.

3.1. Dynamic Property

The dynamic property test consists of dynamic flow test, aeration test, and consolidation test. The torque and axial force on the blade are measured and the resistance to flow is calculated and illustrated as a flow energy. The results of dynamic flow test are shown in Figure 1, including basic flowability energy (BFE), specific energy (SE), stability index (SI), and flow rate index (FRI). The BFE represents the energy required to displace a powder during non-gravitational forced flow, i.e. the powder's resistance to flow in a constrained environment. MetcoAdd 316L powder has a higher BFE which indicates the powder is resistant to the forced flow. The SE measures the energy required for gravity induced flow (resistance to flow in an unconstrained environment). The marginally higher SE value of MetcoAdd 316L powder indicates an increased level of mechanical interlocking and friction. The measurement results of BFE and SE imply the insensitivity of the MetcoAdd 316L powder subjected to powder dispensing of AM process. Powder particles need a higher stimulating energy to flow across the powder bed. The SI is a measure of the physical stability of the material as a function of repeat tests. MetcoAdd 316L powder appears to be physically stable with an SI value close to 1. The higher SI value of FS 316L powder suggests this powder is more sensitive to be handled and processed. The FRI Measures the response of the powder to change flow rates, especially relevant for mixing applications and variable feed rate environments. The FS 316L powder shows a little higher FRI value, indicating a slight sensitivity to changes in flow rate.

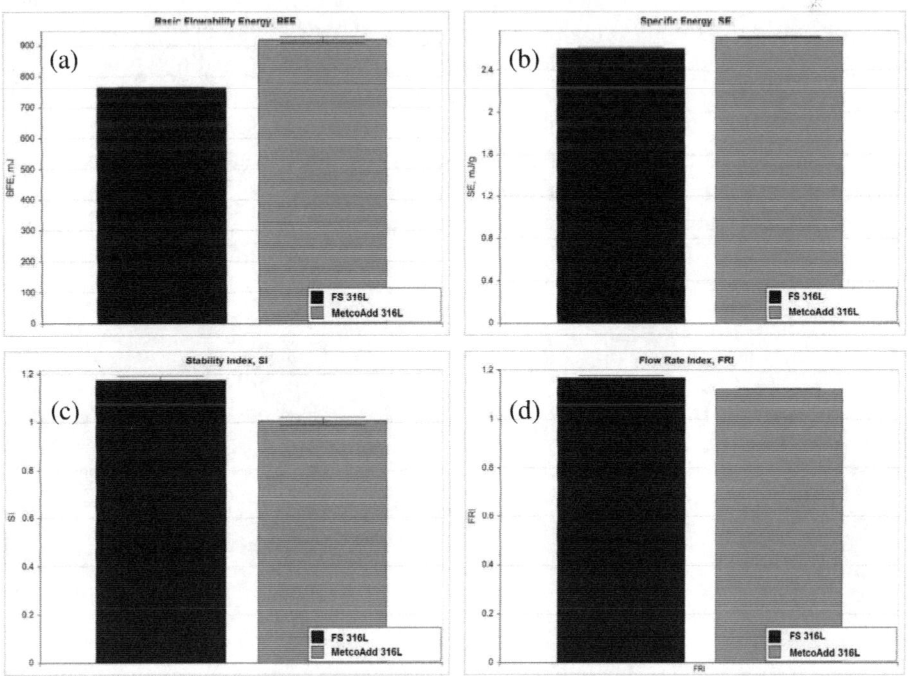

Figure 1. Results of dynamic flow test. (a) basic flowability energy; (b) specific energy; (c) stability index; (d) flow rate index

The aeration test measures the changes in the flow properties due to the introduction of air into the sample. It identifies how powder behaves during mixing, pneumatic conveying and fluidization. It ultimately provides an indication of absolute cohesive strength. As shown in Figure 2, the FS 316L powder presents a higher value of aeration ratio. Thus, this powder is easily to reach a fluidized state at a lower air velocity in comparison with the MetcoAdd 316L powder. The difference of powders to aeration may not be influential to PBF base AM process. But for the LENS process, it could be critical [14]. The consolidation test measures the impact of tapping on the flow properties of a powder. High consolidated energy (CE) value can be indicative of some potential problems during processing or transportation due to consolidation. Figure 3 shows a higher CE value of the MetcoAdd 316L powder. Its particles are more sensitive to vibrational forces and are liable to be packed resulting in increased resistance to flow.

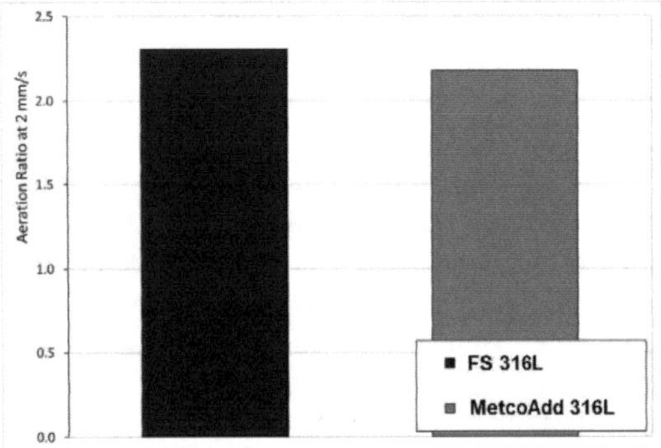

Figure 2. Result of aeration test

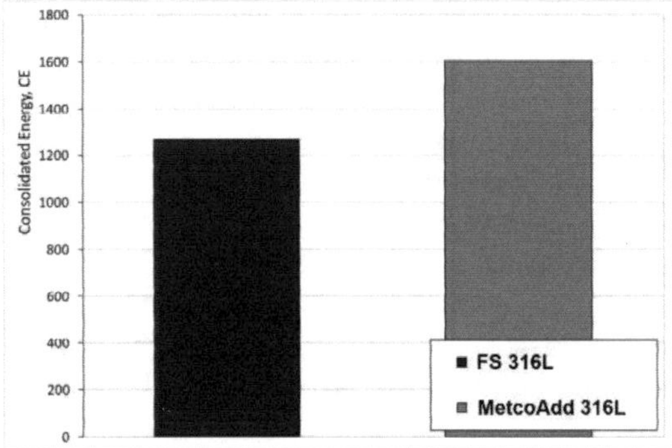

Figure 3. Result of consolidation test

3.2. Bulk Property

Permeability test and compressibility test were conducted to reveal the bulk property of both powders. The permeability test measures the air pressure required to maintain a constant air flow through the powder bed under an increasing normal stress. Generally, a higher normal stress results in a higher Pressure Drop (PD) as the permeability is reduced due to particles becoming more closely packed. MetcoAdd 316L powder exhibits a higher PD value, as shown in Figure 4, indicating a decreased permeability. Low permeability is not desired in the discharging or filling process of metal AM processes. Compressibility is simply a bulk property measurement that reflects the ability of the powder to become consolidated when subjected to a normal stress. It is not a direct measure of flowability but can be used to understand the behavior of volume change when being handled in the powder dispensing process. The FS 316L power and MetcoAdd 316L powder show a comparable compressibility value (Figure 5). Hence, both powders have similar levels of particle packing efficiency.

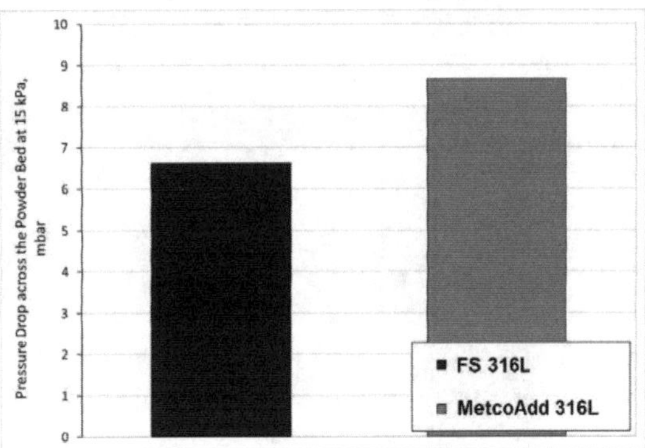

Figure 4. Result of permeability test

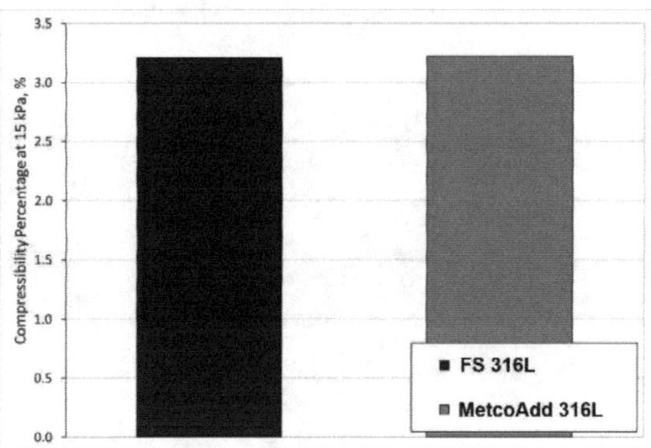

Figure 5. Result of compressibility test

3.3. Shear Property

Shear property measurement includes shear cell test and wall friction test. The shear cell test is intended to determine the shear stress required to initiate flow in a pre-consolidated powder. This provides an indication of how easily a powder will move from a static condition into dynamic flow. Figure 6 shows the result of shear cell test. It is noted that the two powders have comparable shear stress values, indicating a similar resistance to the onset of flow at this level of consolidation. The wall friction test is similar to the shear cell test. Instead of a shear cell, a disc made of a known material is forced to shear against the powder bed surface, rather than the powder being forced to shear against itself. Varied normal stresses are applied to the disc during the test. As shown in Figure 7, the wall friction angle values are also comparable. Hence, both 316L powders exhibit similar shear properties.

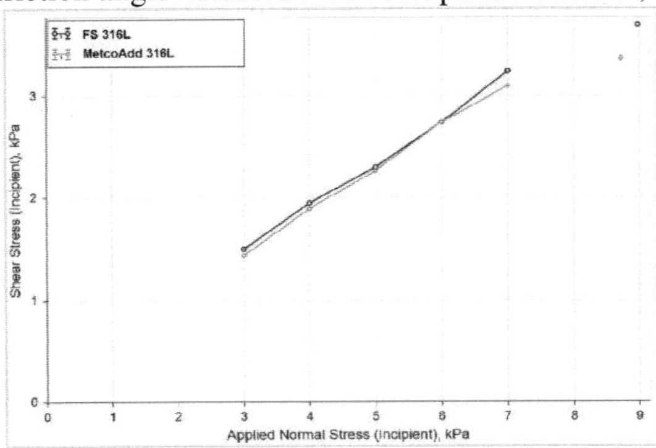

Figure 6. Result of shear cell test

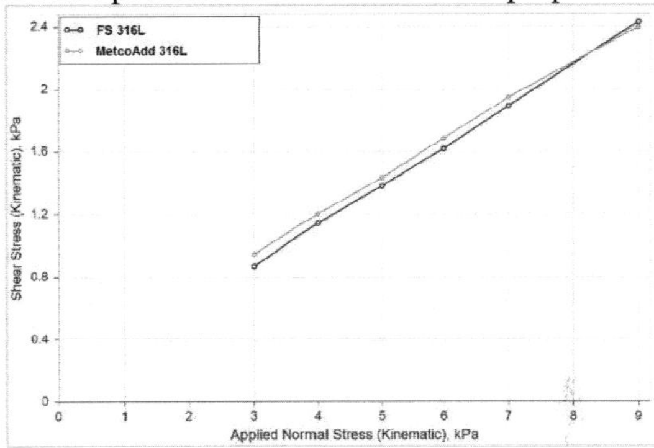

Figure 7. Result of wall friction test

4. Summary

Two stainless steel powders were measured and compared in this study regarding their rheological properties for PBF based AM process. The dynamic property, bulk property, and shear property of both powders were tested and discussed. The dynamic flow test indicates a distinguishable difference of powder flowability. But it is believed that the flow of virgin powder would be improved due to the loss of small particle after several production cycles. Other testing results indicate comparable results of two powders. Overall, it is essential to employ the powder rheology to measure the powder attributes. The testing results confirm that the newly acquired MetcoAdd 316L powder (virgin powder) is compatible to the used FS 316L powder when mixing them.

Reference

[1] Gibson I, Rosen D W, and Stucker B 2015 *Additive Manufacturing Technologies: Rapid Prototyping to Direct Digital Manufacturing* (New York: Springer)
[2] Vock S, Klöden B, Kirchner A, Weißgärber T, and Kieback B *Prog. Addit. Manuf.*, in press.
[3] Gu H, Gong H, Dilip J J S, Pal D, Hicks A, Doak H, and Stucker B 2014 *Solid Freeform Fabrication Symp.* Austin TX, USA
[4] Tang H P, Qian M, Liu N, Zhang X Z, Yang G Y, and Wang J 2015 *JOM* **67** 555
[5] Spierings A B and Levy G 2009 *Solid Freeform Fabrication Symp.* Austin TX, USA
[6] Spierings A B, Herres N, and Levy G 2011 *Rapid Prototyping J.* **17** 195
[7] Slotwinski J A, Garboczi E J, Stutzman P E, Ferraris C F, Watson S S, Peltz M A 2014 *J. Res. Natl. Inst. Stan.* **119** 460
[8] Niu H J, and Chang I T H 1999 *Scripta. Mater.* **41** 25
[9] Spierings A B, Voegtlin M, Bauer T, and Wegener K 2016 *Prog. Addit. Manuf.* **1** 9
[10] Strondl A, Lyckfeldt O, Angré A, and Chasoglou D 2013 *European Powder Metallurgy Congress Proc.* Gothenburg, Sweden
[11] Lyckfeldt O 2013 *European powder metallurgy congress proc.* Gothenburg, Sweden
[12] Clayton J, Millington-Smith D, and Armstrong B 2015 *JOM* **67** 544
[13] Krantz M, Zhang H, and Zhu J 2009 *Powder Technol.* **194** 239
[14] Thornton A, Saad J, Clayton J *Met. Powder Rep.*, in press.

Thermal conductivity improvement of Al alloy with Nb nanopowder using SPS for transmission conductor

C O Ujah[1 3*], **A P Popoola**[1 4], **O M Popoola**[2 5]

[1]Faculty of Engineering and the Built Environment, department of Chemical, Metallurgical and Materials Engineering, Tshwane University of Technology, Pretoria, South Africa

[2] Faculty of Engineering and the Built Environment, department of Electrical Engineering, Tshwane University of Technology, Pretoria, South Africa

[3]omega.ujah@gmail.com, [4]popoolaapi@tut.ac.za, [5]popoolao@tut.ac.za

***Corresponding author**: omega.ujah@gmail.com

Abstract. Aluminium 1000 alloy has high purity and good electrical conductivity required for high power transmission conductors but lacks adequate thermal conductivity which induces sagging in service. To augment this deficiency, Nb nanopowder was incorporated into it as a reinforcement and consolidated with spark plasma sintering (SPS). The SPS was conducted with pre-optimized processing parameters of 630 °C, 30 MPa, 10 mine and 200 °C/min. The sintering was conducted in a vacuum with graphite die of 20 mm in diameter. Thereafter, the sintered samples were metallographically prepared and subjected to thermal conductivity testing. A laser flash analyzer was used to generate the thermal diffusivity of the samples. Then their thermal conductivities were computed. Results obtained showed that Al-4Nb had the highest thermal conductivity of 115. 95 ± 6.9 W/mK which represented 42% improvement. So, it follows that Al-Nb composite is a better replacement for monolithic Aluminium 1000 alloy in power transmission conductors with a potential property of low sag.

Key words: Thermal conductivity; Al alloy; Niobium; High transmission conductor; Laser flash analysis.

1. Introduction

Aluminium 1000 alloy is widely used in power transmission conductors due to its high purity, high electrical conductivity, good passivation property and low density. Neverthless, it has high coefficient of thermal expansion (CTE) due to its poor thermal conductivity. Consequently, it sags excessively when high load is transmitted across it. Meanwhile, researchers have discovered that this challenge can be redeemed by reinforcing it with suitably selected metals, non metals or fibers [1-3]. Moreso, Pietrak and Wisniewski [4] observed that interfacial thermal resistance (ITR) is one of the major causes of drop in thermal conductivity of metal matrix composites (MMCs). This situation arises when matrix/reinforcement interface is loosely bonded; such that flow of electrons are hampered resulting in conductivity drop. Studies have also shown that thermal conductivity of Al matrix composites (AMCs) is dependent on the nature of its reinforcement since it relies on the amount of crystals present in the solid solution as the Guinier Preston (GP) zones/age-hardening precipitates [5]. And that is why Tritt [6] rightly pointed out that solid solution creates the greatest electron and phonon scattering spots on the MMCs. Meanwhile, high power transmission conductors must possess high thermal conductivity so as to contain the expansion and contraction of the conductor during the peak and off-peak hours respectively. So, it is imperative to develop a conductor material with a good thermal behaviour. Interestingly, it has been shown that Nb reinforcement induces the formation of Nb_2O_5 which enhances compression strength on materials [8]. This attribute will control its sag level. Several researchers have used spark plasma sintering (SPS) to compact Al with one or more additives in order to enhance its properties. Guo et al. [5] used CNTs, Wang et al. [6] worked with TiAl-based alloy, Yilmaz [7] utilized Al_2O_3 and SiC while Ujah et al. [8] reinforced with CNTs/Nb. The authors achieved various improvements which were attributed to the SPS technique as well as the properties of the reinforcements used. However, SPS is a non-conventional sintering technique that consolidates powders using pulsed direct current applied concurrently with pressure. It improves the properties of composites, conserves energy, saves time and money, generates cohesive boundaries with refined grains, vaporizes impurities and enhances the dispersion of reinforcements [9, 10]. Likewise, Nb nanopowder is a grain refiner, a toughening and strengthening material and a superconductor; hence, its choice. The excellent characteristics of SPS and Nb manifested in Ujah et al. [3] where the electrical conductivity, microhardness and microstructure of Al alloy were improved considerably. Works aimed at improving the properties of Al alloy with several additives abound in the literature. But little has been discussed on using Nb nanopowder as reinforcement and SPS as fabrication technique to improve the thermal conductivity of Al alloy despite the good characteristics of the material and technique. Therefore, this work was aimed at improving the thermal conductivity of Al using Nb reinforcement and SPS technique with the primary objective of reducing Al alloy's CTE which will subsequently reduce its sagging so as to perform creditably in high transmission conductor.

2. Experimental Procedure

2.1 Materials and method.

The materials (Al 1000 alloy nanopowder and Nb nano powder), the composition, purity and supplier have been reported elsewhere [8]. Accordingly, the powders were mixed vigorously with tubular mixer for 10 hours to ensure homogenous dispersion of reinforcement in the matrix. The appropriate masses were weighed out before the mixing. The masses included that which would produce the following samples, pure Al, Al-1wt.%Nb, Al-4wt.%Nb, Al-8wt.%Nb, Al-12wt.%Nb with the dimensions of 20 mm diameter and 5 mm thickness as shown in Fig. 1b. The blended powders were then weighed into a grahite die and covered with punches as shown in Fig. 1a and placed into the sintering chamber of the SPS. The sintering was performed with SPS machine using optimized parameters of Ujah et al. [9] in a vacuum pressure of 0.605 mbar. Thereafter, the sintered samples where cut into 10 mm x 10 mm x 4 mm dimensions and polished for thermal conductivity testing.

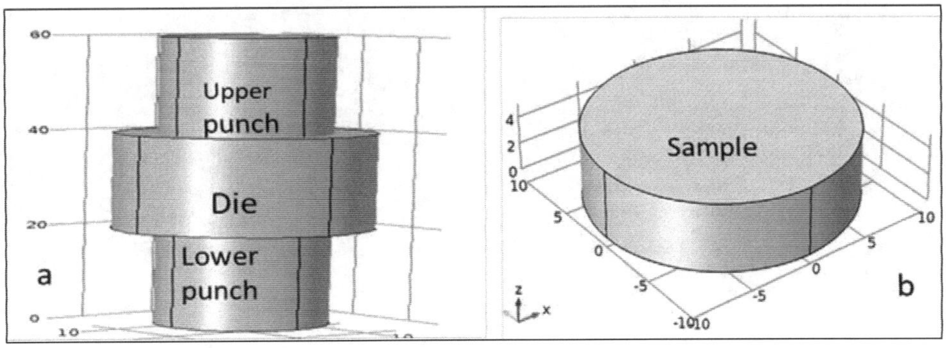

Figure 1. (a) graphite die and punches assembly. (b) Sintered sample

2.2. Microstructural characterization.

As-received powders were analyzed with Transmission electron microscopy (TEM, JEM-2400F), while the sintered samples were studied with scanning electron microscopy connected to electron dispersion spectroscopy (SEM/EDS, FE-SEM, TESCAN).

IOP Conf. Series: Materials Science and Engineering **689** (2019) 012004 doi:10.1088/1757-899X/689/1/012004

2.3. Thermal conductivity testing.

A Laser Flash Analyzer (LFA 427) connected to (NETZSCH Proteus software) was used in this test. The samples were cut to 10 mm by 10 mm dimensions, with a thickness of 4 mm. The samples were polished and coated with graphite to enhance its thermal diffusivity and absorptivity. The test was run in an argon atmosphere, with laser voltage of 450 V and laser pulse width of 0.8 ms. The thermal diffusivity of each sample was measured by sending a laser (energy pulse) to the surface of a sample and generating the time taken for the energy to reach the other side of the sample. Then thermal conductivity was computed with the relationship shown in Eq. 1 as follows:

$$\lambda = \alpha \rho C p \quad (1)$$

Where λ is the thermal conductivity (W/mK), α is the thermal diffusivity (m^2/s), ρ is the density of the sintered material (kg/m^3), Cp is the specific heat capacity of the material (J/KgK).

3. Results and Discussion

3.1. Microstructure.

Fig. 2 shows the TEM micrographs of as-received powders. The grains of Al in Fig. 2a are spherical in shape with tiny particles surrounding large ones. Fig. 2b shows the grains of Nb made of large spherical grains surrounded by smaller particles which form feather-like structure which may aid in bonding to the adjacent grains. Fig. 3 displays SEM micrographs of sintered samples. From the EDS results there in and elsewhere [8], it could be seen that the white patches are Nb grains. In Fig. 3a, the white particles were not seen because it was a sample of pure Al which equally confirms that assertion. Al-1Nb SEM micrograph shown in Fig. 3b has the white grains of Nb dispersed scantly in the microstructure. This shows that 1 wt. % was not enough in microstructure. However, in Fig. 3c, the percentage of Nb was 4 wt. % and could be seen to have dispersed evenly to the entire surface. Also, it could be seen in Fig. 3d (Al-8Nb) that there was a very high concentration of Nb as depicted by the white reflective particles. More so, in Fig. 3e (Al-12Nb), the white particles became so much pronounced on the surface. In general, the grains of the sintered samples (Fig. 3) can be seen to be more refined, spherical and closely packed which were as a result of the sintering [10] and this shows that the SPS consolidation enhanced densification and grain refinement in Al-Nb composites. There were no pores and the microstructure was devoid of impurities as was reported by Xie et al. [10], and there was homogenous dispersion of reinforcements. The grain boundaries were tightly bonded, and this was also observed by Ujah et al. [9].

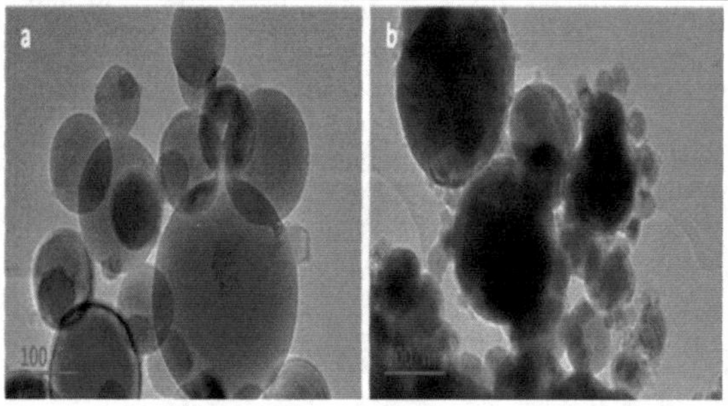

Figure 2. TEM images of as-received powders

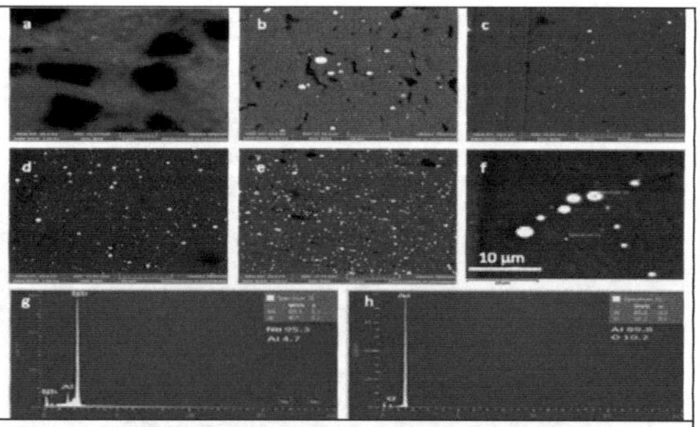

Figure 3. SEM images of sintered samples

3.2 Thermal conductivity

Fig. 4 shows the thermal conductivity plot of sinetered samples. In Fig. 4a, it could be seen that the the thermal conductivity decreased with increase in temperature. This would be attributed to the thermal conductivity drop induced by electron and phonon scattering phenomenon when heat flow is increased in a system as reported by Baudouy and Four [11]. However, the addition of 1 wt. % Nb increased the average thermal conductivity of pure Al from 81.55 ± 9.0 W/mK to 108.05 ± 7.6 W/mK as can be seen in Fig. 4b. Again, when the percentage weight of Nb was increased to 4, the thermal conductivity increased to 115.95 ± 6.9 W/mK. But when the increase got to 8 wt.%Nb, the thermal conductivity decreased to 99.13 ± 8.0 W/mK. At 12 wt. % Nb, the thermal conductivity decreased even lower than the pure Al, it became 74.38 ± 8.0 W/mK. The improvement was effective when the Nb content was within the range of 1 to 4 wt.%. The improvement could be attributed to grain refining property of Nb with its superconductivity and the exploits of SPS technique which impacted excellent microstructure on the composites as observed in Al-1Nb and Al-4Nb samples. Also their microstructures attested to their superior properties because Fig. 3b and 3c show microstructure without pores, with refined grains and compact interface. Hence, the samples had cohesively bonded grain boundaries which induced free flow of electrons and enhance thermal conductivity which is ususally absent in weakly bonded matrix/reinforcement interface as observed by Krishna et al. [12]. Meanwhile, the decline in thermal conductivity in Al-8Nb and Al-12Nb nanocomposites would be attributed to the fact that the weight fractions of Nb seemed to have exceeded optimal concentration required for thorough wetting out. For this reason, they got agglomerated in the grain boundaries as can be seen in Fig. 3e which rather induced thermal resistivity. Therefore, 4 wt.% Nb was the optimum concentration that gave the highest thermal conductivity of 42% improvement.

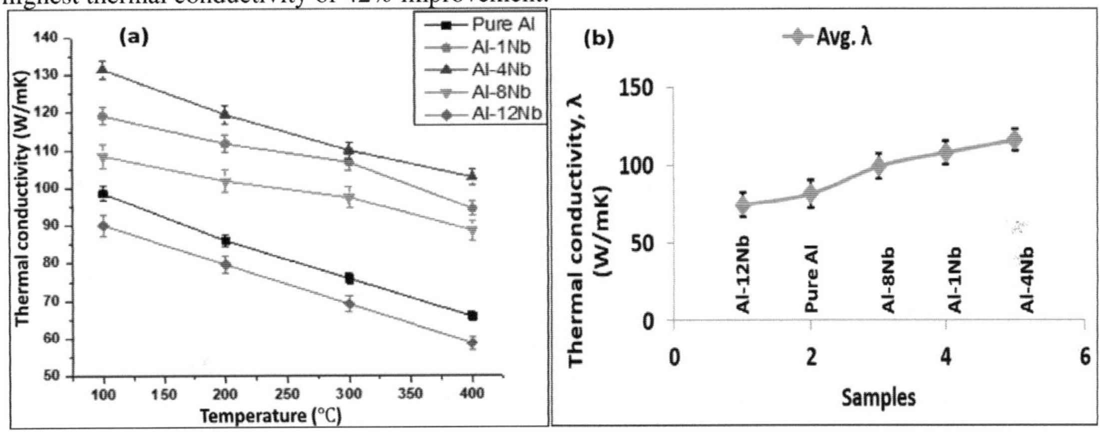

Figure 4. (a) Thermal conductivity of sintered samples. (b) Average thermal conductivities

4. Conclusion

The study aimed at enhancing the thermal conductivity of Al 1000 alloy was successfully conducted and the results obtained showed that Al-4Nb composite got improved in thermal conductivity by 42%. This was attributed to the enhanced properties of the developed sample with microstructure devoid of pores, and exhibited high grain refinement with cohesively bonded boundaries. And all these attributes were impacted by the SPS fabrication technique and the grain refining property Nb reinforcement. However, Al-12Nb sample had poor wetting out of the reinforcement which resulted in its agglomeration in the boundaries and a subsequent drop in the thermal conductivity of the sample. Therefore, it could be said that the optimum weight percentage of Nb adequate to improve the thermal conductivity of Al is 4 wt. %. So, it can be concluded that Al-4Nb nanocomposite would perform much better than monolithic Al alloy in terms of thermal conductivity in power transmission conductors.

5. Acknowledgement

The authors want to appreciate centre for energy and electric power (CEEP) and Tshwane University of Technology (TUT) Pretoria for their financial assistance in the course of this work. There was no conflict of interest in this work.

References

[1] Alawar A, Bosze E J and Nutt S R 2005 *A composite core conductor for low sag at high temperatures.* IEEE transactions on power delivery, **20**(3): p. 2193-2199.
[2] Banerjee K 2014 *Making the case for high temperature low sag (htls) overhead transmission line conductors.* Arizona State University.

[3] Ujah C O, Popoola A P, Popoola OM and Aigbodion V 2019 *Electrical conductivity, mechanical strength and corrosion characteristics of spark plasma sintered Al-Nb nanocomposite.* The International Journal of Advanced Manufacturing Technology, p. 1-8.

[4] Pietrak K and Wiśniewski T S 2014 *A review of models for effective thermal conductivity of composite materials.* Journal of Power Technologies, **95**(1): p. 14-24.

[5] Guo B, et al 2017 *Improving the mechanical properties of carbon nanotubes reinforced pure aluminum matrix composites by achieving non-equilibrium interface.* Materials & Design, **120**: p. 56-65.

[6] Wang D, Yuan H, and Qiang J 2017 *The microstructure evolution, mechanical properties and densification mechanism of TiAl-based alloys prepared by spark plasma sintering.* Metals, **7**(6): p. 201.

[7] Yilmaz S O 2007 *Comparison on abrasive wear of SiCrFe, CrFeC and Al2O3 reinforced Al2024 MMCs.* Tribology international, **40**(3): p. 441-452.

[8] Ujah C O, Popoola A P, Popoola O M and Aigbodion V S 2019 *Enhanced mechanical, electrical and corrosion characteristics of Al-CNTs-Nbcomposite processed via spark plasma sintering for conductor core.* Journal of Composite Materials, p. 0021998319848055.

[9] Ujah, C O, Popoola O M and Aigbodion V S 2019 *Optimisation of spark plasma sintering parameters of Al-CNTs-Nb nano-composite using Taguchi Design of Experiment.* The International Journal of Advanced Manufacturing Technology, 2019. **100**(5-8): p. 1563-1573.

[10] Xie G, et al. 2003 *Behavior of oxide film at the interface between particles in sintered Al powders bypulse electric-current sintering.* Metallurgical and Materials Transactions A, **34**(3): p. 699-703.

[11] Baudouy B and Four A 2014 *Low temperature thermal conductivity of aluminum alloy 5056.* Cryogenics, **60**: p. 1-4.

[12] Krishna S M, Shridhar T and L. Krishnamurthy L 2015 *Experimental Investigations on Thermal Analysis and Thermal Characterization of Al 6061-Sic-Gr Hybrid Metal Matrix Composites.* International Journal of Material Science, **5**(2): p. 54.

Influence of varied process parameters on the microstructure, densification and microhardness of spark plasma sintered Ti-6Al-4V/h-BN binary composite

J O Abe[1,2], A P I Popoola[1] and O M Popoola[2]

[1] Department of Chemical, Metallurgical and Materials Engineering, Tshwane University of Technology, P.M.B. X680, Pretoria, South Africa
[2] Centre for Energy and Electric Power, Tshwane University of Technology, Pretoria, South Africa

Email: johnabe.fm@gmail.com

Abstract. The influence of varied process parameters involving three levels of temperature, pressure, heating rate and dwell time on the microstructure, densification and microhardness of spark plasma sintered Ti-6Al-4V/*h*-BN binary composite was studied. Taguchi design method was adopted to randomize the SPS process parameter levels. The microstructure was analyzed by optical microscopy, densification was determined based on the principle of Archimedes and Vickers microhardness tester was used to evaluate the microhardness of the sintered composite. The consolidation of Ti-6Al-4V powder and nanoparticles of 3 wt.% of *h*-BN via SPS produced nearly full theoretical densification of 99.77% and the sintered composite gave a microhardness value of 710.37 HV which is more than 200% that of the monolithic alloy. The best combination of relative densification and microhardness were obtained at the sintering parameters of 1000 °C of temperature, 40 MPa of pressure, 100 °C/min heating rate and 15 min of dwell time. Generally, it was found that improved microstructure, densification and microhardness were influenced by the high temperature of sintering and low rate of heating that gave room for adequate diffusional mass transport resulting in solid bonding between the particles of the matrix and the ceramic reinforcement.

1. Introduction

Ti–6Al–4V alloy combines high strength-to-weight ratio, excellent corrosion resistance, good high-temperature strength, stiffness, good toughness, moderate density, low modulus and good biocompatibility. Thus, the alloy gains enormous attention in the aerospace, chemical and petrochemical, marine, energy and biomaterials industries. Nevertheless, the alloy still suffers from inadequate wear, hardness and poor antifriction properties demonstrated in certain service conditions which have limited its general usage [1-2]. Recently, titanium matrix composites (TMCs) have been developed which possess high strength, good wear resistance, high-temperature stability owing to the added reinforcement. Most sought after set of mechanical properties are commonly realized when fine and thermally stable ceramic particulates are added to a Titanium Matrix Composite [3-4].

Hexagonal boron nitride (*h*-BN) ceramic is the most stable BN phase widely employed in various field of aerospace, high temperature technology, electrotechnology, metallurgical and chemical engineering applications [5-6] due to its excellent lubrication properties, low density, high temperature stability, chemical inertness, high thermal conductivity and non-toxicity [7-8]. In recent years, many studies showed creditable mechanical, chemical, lubricating and high-temperature properties of hexagonal boron nitride [9-10]. Consequently, *h*-BN has been conceived to be an effective reinforcement to compensate for the present shortcomings of Ti-6Al-4V alloy. However, due to the poor sinterability of *h*-BN, it is continuously difficult to manufacture its dense products by using conventional consolidation methods [9-11]. Yang et al. [11] reported successful densification of hexagonal boron nitride through self-densifying mechanism via spark plasma sintering (SPS). SPS can process advanced nanomaterials with controlled grain growth, rapid heating/cooling rates, reduced sintering times, enhanced microstructure, better densification combined with improved mechanical properties [12-13]. Moreover, the production technology of hexagonal boron nitride–reinforced Ti–6Al–4V alloy has hardly ever been reported.

A major concern in a production line is the quality of products and the success of an industry is determined by the consistency and high production rate of quality products. The process parameters and the quality of the products are well connected. Inappropriate combination of process parameters can cause defects in products and optimal combination of process parameters can improve the products quality meaningfully [14,15]. The basic SPS process parameters comprise the temperature, pressure, heating rate and dwell time. These parameters are key to the quality of the sintered products. Hence, it is imperative to randomize these parameters to obtain optimal sintered product quality characteristics and maximize scarce resources concurrently. A very useful method employed in the randomization process is Taguchi method. Taguchi method is a simple, economical, quick and feasible method in designing high-quality processes and in recognizing proper process parameters to achieve optimal results [14,16-17].

In this work, the authors studied the influence of varied process parameters involving three levels of temperature, pressure, heating rate and dwell time on the microstructure, densification and microhardness of spark plasma sintered Ti-6Al-4V/*h*-BN binary composite. Taguchi design method was adopted to randomize the SPS process parameter levels.

2. Materials and methods

Commercially pure Ti-6Al-4V powder (>99%, 45-90 μm spherical particle sizes by TLS Technik, Germany) and *h*-BN (>99%, 100-200 nm hexagonal crystal form by Hongwu International Group Ltd., China) were used as the matrix and reinforcements respectively. The powders were measured according to the stoichiometric proportion of Ti6Al4V-3% *h*-BN with a high precision scale and mixed in a tubular mixer rotating at a speed of 100 rpm for 10 hours to achieve homogeneity.

In the Taguchi design, four parameters: temperature, pressure, heating rate and dwell time, were chosen at three levels: A (800°C, 900°C and 1000°C), B (30MPa, 40MPa and 50MPa), C (100°C/min, 300°C/min and 300°C/min) and D (5mins, 10mins and 15mins). Minitab 19 was used to generate Taguchi L9 3^4 orthogonal array with four columns and nine rows, indicating a total of nine runs of experiment as shown in table 1 where A represents the temperature in °C, B represents the pressure in MPa, C represents the heating rate in °C/min, and D represents the dwell time in min. Correspondingly, 1 represents the least, 2 represents the average, and 3 represents the greatest of the parameters.

The admixed powders were measured in a graphite die of ϕ30mm to produce 10mm thickness of the composite and consolidated according to the design in Table 1 by using spark plasma sintering machine (SPS FCT Systeme GmbH, Germany) under argon atmosphere. The density measurements of the sintered samples were carried out by using Archimedes' principle where the average value of six repeated measurements was taken for each sample and recorded as the effective density and the theoretical density was calculated according to the rule of mixture. Subsequently, the relative densification of each sample was calculated as the ratio of the effective density and theoretical density multiplied by 100%.

Later, the samples were first grinded by using 120-4000 grit papers and then polished by using diamax powder (6 μm, 3 μm and 1 μm) in solution to achieve mirror-like polish. Afterward, Kroll's reagent was used to etch the samples, and the surfaces were observed under an optical microscope.
Finally, microindentation of the sintered samples was done by using Vickers microhardness tester (FUTURE-TECH FM-800 from Japan) with a dura scan diamond indenter and test load of 300gf for 15 seconds with a spacing of 0.5. The average value of five indentations was taken for each sample and recorded as the effective microhardness value of the samples.

Table 1. Taguchi L9 3^4 orthogonal array generated for the experiment and symbols.

Experiment run	SPS parameters and levels				Symbol
	A	B	C	D	
1	1	1	1	1	A1B1C1D1
2	1	2	2	2	A1B2C2D2
3	1	3	3	3	A1B3C3D3
4	2	1	2	3	A2B2C3D1
5	2	2	3	1	A2B1C2D3
6	2	3	1	2	A2B3C1D2
7	3	1	3	2	A3B1C3D2
8	3	2	1	3	A3B2C1D3
9	3	3	2	1	A3B3C2D1

A = temperature in °C, B = pressure in MPa, C = heating rate in °C/min, and D = dwell time in min. 1 = least, 2 = average, and 3 = greatest of the parameters.

3. Results and discussion

The details of the results of the experiment obtained for relative densification and microhardness measurements and microstructure study is seen in table 2 and figure 1, while the individual effect of temperature, pressure, heating rate and dwell time on relative densification and microhardness results was evaluated by calculating the arithmetic means of all the relative densification and microhardness values for the same category of levels of the parameters as shown in table 3 and figures 2-5 respectively.

The optical micrographs of the sintered composite at various sintering parameters are observed in figure 1 (a-i) in order to reveal the dispersion of *h*-BN nanoparticles and its influence on the densification and hardness property of the matrix. The relative densification of the spark plasma sintered monolithic alloy calculated from an effective density of 4.42 g/cm^3 and theoretical density of 4.50 g/cm^3 and Vickers microhardness were 98.22% and 331.8 HV respectively. Among the experimental runs of varied parameters and levels in table 2, A3B2C1D3 gave the best quality characteristics in terms of highest relative densification of 99.77% and microhardness of 710.4 HV.

Table 2. Densification and microhardness measurements results.

Experimental run	SPS parameters and levels	Effective density (g/cm^3)	Theoretical density (g/cm^3)	Relative densification (%)	Microhardness (HV0.3/15)
1	A1B1C1D1	4.33	4.38	98.86	614.9
2	A1B2C2D2	4.30	4.38	98.17	623.2
3	A1B3C3D3	4.14	4.38	94.52	467.0
4	A2B2C3D1	4.28	4.38	97.72	519.1
5	A2B1C2D3	4.25	4.38	97.03	479.3
6	A2B3C1D2	4.31	4.38	98.40	630.1
7	A3B1C3D2	4.34	4.38	99.09	702.6
8	A3B2C1D3	4.37	4.38	99.77	710.4
9	A3B3C2D1	4.32	4.38	98.63	662.8

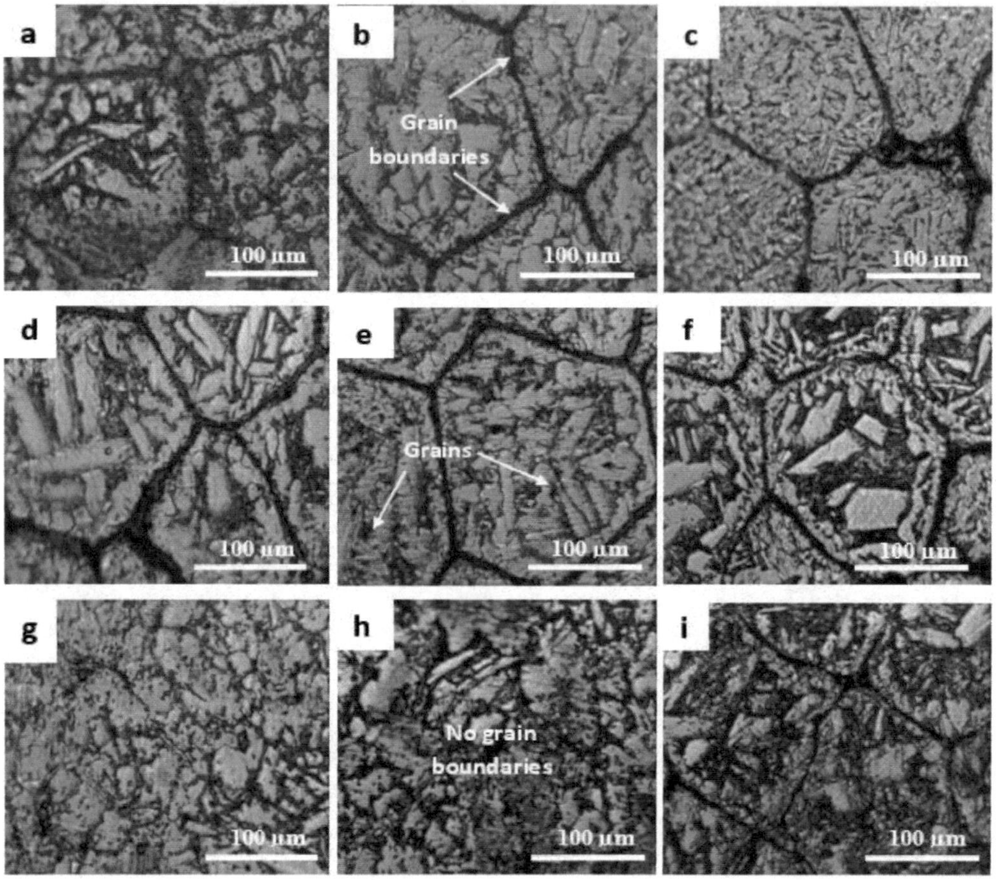

Figure 1. Optical micrographs of the sintered composite at varied process parameters and levels: a. A1B1C1D1, b. A1B2C2D2 c. A1B3C3D3 d. A2B1C2D3 e. A2B2C3D1 f. A2B3C1D2 g. A3B1C3D2 h. A3B2C1D3 i. A3B3C2D1.

Table 3. Individual influence of temperature, pressure, heating rate and dwell time on relative densification and microhardness results.

Parameter	Temperature (°C)			Pressure (MPa)			Heating rate (°C/min)			Dwell time (min)		
Level	800	900	**1000**	**30**	40	50	**100**	200	300	5	**10**	15
Mean relative densification (%)	97.18	97.71	**99.16**	**98.56**	98.32	97.18	**99.01**	98.17	96.88	98.17	**98.55**	97.34
Mean microhardness (HV)	568.4	542.8	**691.9**	**612.2**	604.3	586.6	**651.8**	601.7	549.6	568.6	**652.0**	565.5

Optimum levels of process parameters and results are designated by bold values.

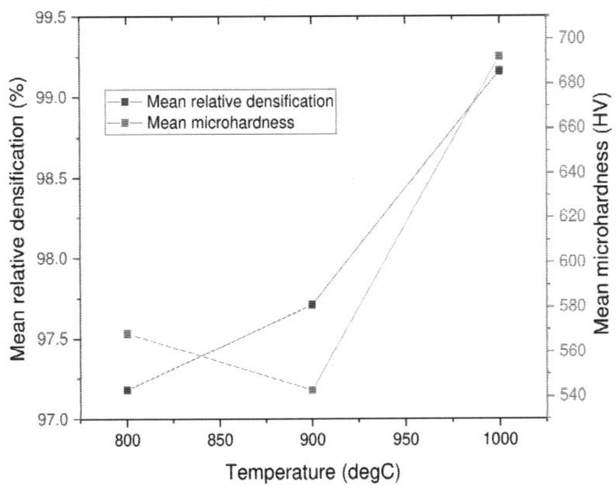

Figure 2. Plot of temperature against mean relative densification and mean.

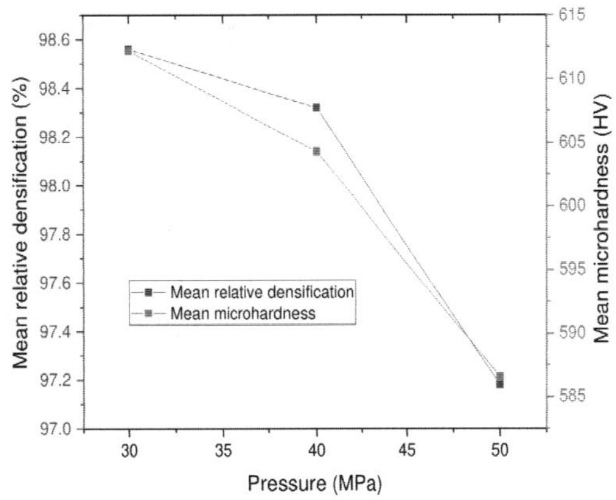

Figure 3. Plot of pressure against mean relative densification and mean microhardness values.

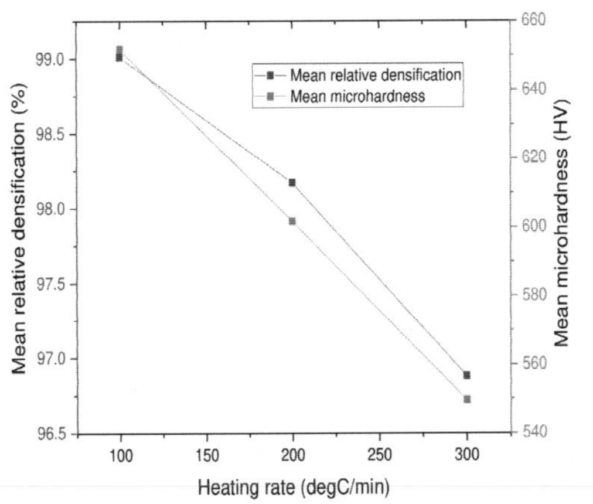

Figure 4. Plot of heating rate against mean relative densification and mean microhardness.

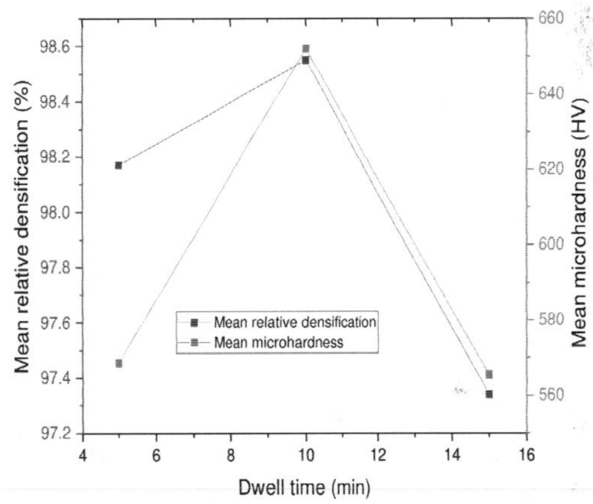

Figure 5. Plot of dwell time against mean relative densification and mean microhardness.

This can be attributed to adequate mixing of the powders where the nanoparticles of *h*-BN filled up pore spaces within the matrix which are the primary causes of poor densification and the effectiveness of the consolidation method which enables the development of dense products [18] as seen in Figure 1h (A3B2C1D3). In addition, there is an indication of a slight grain refinement and disappearance of pores and grain boundaries which explain the reason for the complete densification [17] in contrast to Figure 1c (A1B3C3D3) where the porosity is highest (5.48%) which is significant enough to affect the material properties and led to low relative densification of 94.52% and microhardness of 467 HV. In general, it is realized that improved microstructure, densification and microhardness are promoted by the high temperature of sintering and low rate of heating that gave room for adequate diffusional mass transport resulting in solid bonding between the particles of the matrix and the ceramic reinforcement. Moreover, the significant improvement in the hardness property of the produced composite suggests that SPS facilitates the production of ceramic reinforced metal matrix composites with superior characteristics useful for many engineering applications [18].

From table 3 and figures 2-5, it can be seen that an increase in relative densification and microhardness was noticed when temperature was raised from 800 to 1000 °C and dwell time was raised from 5 to 10 min, whereas a general reduction in relative densification and microhardness was noticed when pressure, heating rate and dwell time were raised from 30 to 50 MPa, 100 to 300 °C/min and 10 to 15 min respectively. A critical consideration of the various results obtained from assessing the individual influence of each parameter on the relative densification and microhardness of the sintered composite as presented in table 3 and figures 2-5 gives the needed information on the most important parameter levels that gave the optimum combination of quality characteristics in the experiment. Hence, the optimum process parameter levels are determined to be the temperature of 1000 °C, pressure of 30 MPa, heating rate of 100°C/min, and dwell time of 10 mins (A3B1C1D2). This suggests that an inappropriate combination of the parameters can potentially lead to a very poor-quality products [15-16] as seen in A1B3C3D3 with 94.52% and 467 HV and A2B2C3D1 with 97.03% and 479.3 HV relative densification and microhardness respectively; in contrast to other improved product qualities with the best being A3B2C1D3 with relative densification of 99.77% and microhardness of 710.4 HV.

4. Conclusion

1. The consolidation of Ti6Al4V powder and nanoparticles of 3wt.% of *h*-BN via SPS produced nearly full theoretical densification of 99.77% and the sintered composite gave a microhardness value of 710.4 HV which is more than 200% that of the monolithic alloy.
2. An increase in relative densification and microhardness was noticed when temperature was raised from 800 °C to 1000 °C, whereas a general reduction in relative densification and microhardness was noticed when pressure, heating rate and dwell time was raised from 30 to 50 MPa, 100 to 300 °C/min and 10 to 15 min respectively.
3. According to the experiment, the best combination of relative densification and microhardness were obtained at the sintering parameters of 1000 °C of temperature, 40 MPa of pressure, 100 °C/min of heating rate and 15 min of dwell time, whereas results show that the optimal parameter levels were obtained at 1000 °C of temperature, 30 MPa of pressure, 100 °C/min heating rate and 10 min of dwell time.
4. Generally, it is found that improved microstructure, densification and microhardness were influenced mostly by the high sintering temperature and low heating rate.

5. Acknowledgment

The authors would like to acknowledge the support received from the Department of Chemical, Metallurgical and Materials Engineering and Centre for Energy and Electric Power (CEEP), Tshwane University of Technology (TUT), Pretoria, South Africa.

References

[1] Ochonogor, O F, Akinlabi, E T, Nyembweb, K D, Pityana, S, & Shongwe, M B (2017). Particle variation and effect on the microstructure and microhardness of Ti6Al4V hybrid metal matric system. *International conference on sustainable materials processing and manufacturing, SMPM* (pp. 616-621). Kruger National Park: Procedia manufacturing.

[2] Mouritz, A. P. (2012). *Introduction to aerospace materials.* Philadelphia New Delhi: Woodhead Publishing Limited.

[3] Kim, I. Y., Choi, B. J., Kim, Y. J., & Lee, Y. Z. (2011). Friction and wear behaviour of titanium matrix (TiB+TiC) composites. *Wear 271*, 1962-1965.

[4] Ray, A. K., Venkateswarlu, K., Chaudhury, S. K., Das, S. K., & Kumar, B. R. (2002). Fabrication of TiN reinforced aluminium metal matrix composites through a powder metallurgical route. *Materials Science and Engineering A338*, 160-165.

[5] Jiang, X., Weng, Q., Wang, X., Li, X., Zhang, J., Golberg, D., & Bando, Y. (2015). Recent Progress on Fabrications and Applications of Boron Nitride Nanomaterials: A Review. *Journal of Materials Science & Technology 31*, 589-598.

[6] Ertuğ, B. (2013). In *Powder Preparation, Properties and Industrial Applications of Hexagonal Boron Nitride* (pp. 33-56). IntechOpen.

[7] Lipp, A., Schwetz, K. A., & Hunold, K. (1989). Hexagonal boron nitride: fabrication, properties

IOP Conf. Series: Materials Science and Engineering **689** (2019) 012005 doi:10.1088/1757-899X/689/1/012005

and applications. *Journal of the European Ceramic Society, 5*, 3-9.

[8] Eichler, J., & Christoph, L. (2008). Boron nitride (BN) and BN composites for high-temperature applications. *Journal of the European Ceramic Society 28*, 1105–1109.

[9] Du, Y., Li, S., Zhang, K., & K, L. (1997). BN/Al composite formation by high-energy ball milling. *Scripta Materialia, 36(1)*, 7-14.

[10] Firestein, K. L., Steinman, A. E., Golovin, I. S., Cifre, J., Obraztsova, E. A., Matveev, A. T., Golberg, D. (2015). Fabrication, characterization, and mechanical properties of spark plasma sintered Al–BN nanoparticle composites. *Materials Science & Engineering A 642*, 104–112.

[11] Yang, H; Fang, H; Yu, H; Chen, Y; Wang, L; Jiang, W; Wu, Y; Li, J. (2019). Low temperature self-densification of high strength bulk hexagonal boron nitride. *Nature Communications 10*, 1-9.

[12] Al-Aqeeli, N. (2013). Processing of CNTs Reinforced Al-Based Nanocomposites Using Different Consolidation Techniques. *Journal of Nanomaterials*, 1-10.

[13] Diouf, S., & Molinari, A. (2012). Densification mechanisms in spark plasma sintering: Effect of particle size and pressure. *Powder Technology 221*, 220–227.

[14] Matizamhuka, W. R. (2016). Spark plasma sintering (SPS) – an advanced sintering technique for structural nanocomposite materials. *The Journal of the South African Institute of Mining and Metallurgy Volume 16*, 1171-1180.

[15] Fei, N. C., Mehat, N. M., & Kamaruddin, S. (2013). Practical Applications of Taguchi Method for Optimization of Processing Parameters for Plastic Injection Moulding: A Retrospective Review. *ISRN Industrial Engineering*, 1-11.

[16] Athreya, S., & Venkatesh, Y. D. (2012). Application of Taguchi Method for Optimization of Process Parameters In Improving The Surface Roughness of Lathe Facing Operation. *International Refereed Journal of Engineering and Science, 1(3)*, 2319-1821.

[17] Ujah, C. O., Popoola, A. P., Popoola, O. M., & Aigbodion, V. S. (2018). Optimisation of spark plasma sintering parameters of Al-CNTs-Nb nano-composite using Taguchi Design of Experiment. *The International Journal of Advanced Manufacturing Technology*, 1-11.

[18] Cavaliere, P., 2019. *Spark Plasma Sintering of Materials: Advances in Processing and Applications.* 1st ed. Switzerland: Springer Nature.

Wear and surface roughness analysis of machining of Ti-6Al-4V under dry, wet and cryogenic conditions

Muhammad Ali Khan[1,a], Syed Husain Imran Jaffery[1,b], Mushtaq Khan[1,c] and Shahid Ikramullah Butt[1,d]

[1]Department of Design and Manufacturing Engineering (DME), School of Mechanical and Manufacturing Engineering (SMME), National University of Sciences and Technology (NUST), Sector H-12, 44000, Islamabad, Pakistan

Email: [a]muhammadali.pg@smme.edu.pk, [b]imran@smme.edu.pk, [c]mkhan@smme.edu.pk, [d]drshahid@smme.nust.edu.pk

Abstract. Use of cooling media contributes in elevation of desired responses owing to its cooling and lubrication effects especially in case of hard to cut materials like Ti-6Al-4V. This study aims to analyze the effect of cooling media in addition to cutting speed on responses including tool wear and surface integrity during turning. Results highlighted the effectiveness of cooling system in particular that of cryogenic media. Use of cryogenic media and cutting fluid enhanced tool life max by 33% and 21% respectively. At lower cutting speed surface roughness was improved by 13% and 8% by cryogenic media and cutting fluid respectively.

Keywords: Ti-6Al-4V, Titanium, Machining, Cryogenic

1. Introduction

Use of cooling media is as old as the history of machining itself. From basic water cooled machining systems to state of the art cryogenic cooling, technology has evolved into different forms in pursuit of overall process optimization. Selection of cooling media for a specific manufacturing system merits necessary consideration as it accounts for about 15% of the total manufacturing costs [1]. Minimum quantity lubrication [2] and conventional high volume coolant [3] are widely used in different machining systems. Lately cryogenic media has served the purpose well by reducing the cutting zone temperature and lubricating the sliding surfaces reducing tool wear and surface roughness respectively. Added advantages include no extra disposal measures and ease of global availability [4].

Economy and productivity defines efficiency of any manufacturing system [5]. Prolonged tool life contributes towards economy whereas better surface finish improves product quality. Ti-6Al-4V is a widely used titanium alloy because of good corrosion resistance, high strength and hardness. Approximately 60% of all titanium alloys usage consist of Ti-6Al-4V [6]. On the other hand it is highly reactive at high temperatures associated with high cutting speeds [7] in addition to its low thermal conductivity. Low modulus of elasticity and high temperature strength also adversely effects tool life [8]. Tooling costs make up around 4% [1]of total cost which is a sizeable figure keeping in view the overall manufacturing volume. Surface integrity also holds position of central importance as any product not conforming to requirement will amount to waste and add up to inefficiency. Working with Ti-6Al-4V, Jerold and Kumar [9] found that liquid nitrogen (LN2) and carbon dioxide (CO2) both produced better surface finish and tool life than dry machining. Sun et al [10]reported improved tool life turning Ti-6Al-4V under cryogenic conditions owing to reduced cutting zone temperature and shorter chip contact length. Cryogenic conditions also gave better tool life and improved surface roughness in milling operations [11]. Birmingham et al [12]concluded in his research that high pressure coolant and LN2 produced comparable results in term of tool wear which were much better than dry machining. Keeping in view the effectiveness of both wet and cryogenic machining present research was conducted by vestingthe cutting condition (dry, wet and cryogenic) as an input variable with varying cutting speed.

2. Experimental setup

CNC turning centre (ML 300) was used in experiments as shown in Fig. 1. Cryogenic LN2 setup was established using cryogenic cylinder with 160 liters capacity. Efficient dual jet configuration [13] was used connected via cryogenic bifurcating valve and decanting pipe to the cylinder. Temperature around -196°C is reached with use of LN2 cryogenic media [14]. Ti-6Al-4V was selected as work piece with chemical composition as shown in Table 1. ML 300 internal cooling system was used for wet machining. Dry, wet and cryogenic experiments were carried out at each of the four conditions given in Table 2.

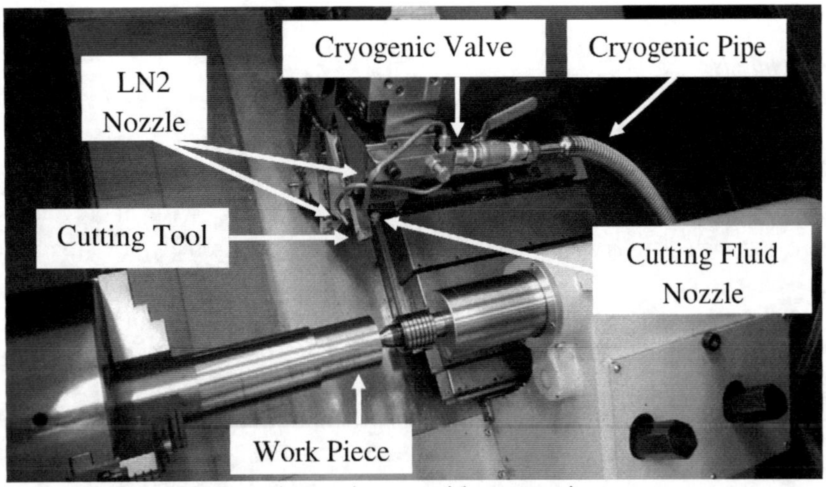

Figure 1. Experimental setup with cryogenic arrangement

Table 1. Chemical Composition (wt. %)

Ti	V	Al	Fe	Cu	Cr
89.44	4.2	5.7	0.15	0.003	0.0023

Table 2. Design of experiment

Condition	Inputs		
	Feed rate (f) (mm/rev)	Cutting speed (v) (m/min)	Depth of cut (d) (mm)
1	0.12	50	1
2	0.12	75	1
3	0.12	100	1
4	0.12	125	1

3. Response measurement and analysis

Tool wear rate was measured using flank wear by Eq. 1[15]which is an appropriate comparison as shown in Table 3. Higher negative value represents lower flank wear and vice versa. Surface roughness results as tabulated in Table 3 were determined using roughness tester TR 110. These results were further evaluated and compared as shown in Fig. 2.

$$R = log\left[\frac{VB}{l_s}\right] = \left[\frac{VB}{1000tV_c}\right] \quad (1)$$

Table 3. Tool wear rate and surface roughness results

Condition	Tool wear rate			Surface roughness		
	Dry	Wet	Cryogenic	Dry	Wet	Cryogenic
1	-6.15	-6.25	-6.32	1.55	1.44	1.36
2	-6.09	-6.16	-6.20	1.39	1.30	1.24
3	-6.01	-6.06	-6.09	1.28	1.23	1.16
4	-5.98	-6.01	-6.04	1.11	1.08	1.03

3.1 Tool wear analysis

Tool wear has a relationship of direct proportionality with cutting speed as evident in Table 3. As temperature acts as a catalyst in wear inducing mechanisms [16] tool wear increases with increasing speed as cutting zone temperature elevates as speed increases [17]. Use of cooling media has significantly reduced tool wear as highlighted in Fig. 2 (a) with tool wear improvement indicated at each condition. In particular, cryogenic media because of its extremely low temperature ensured reduced heat production as well as timely heat extrication resulting in improved tool life. It was observed that effect of cooling media abridged at speeds above 100 m/min because of the reduced penetration in the cutting zone at such higher speeds [18].

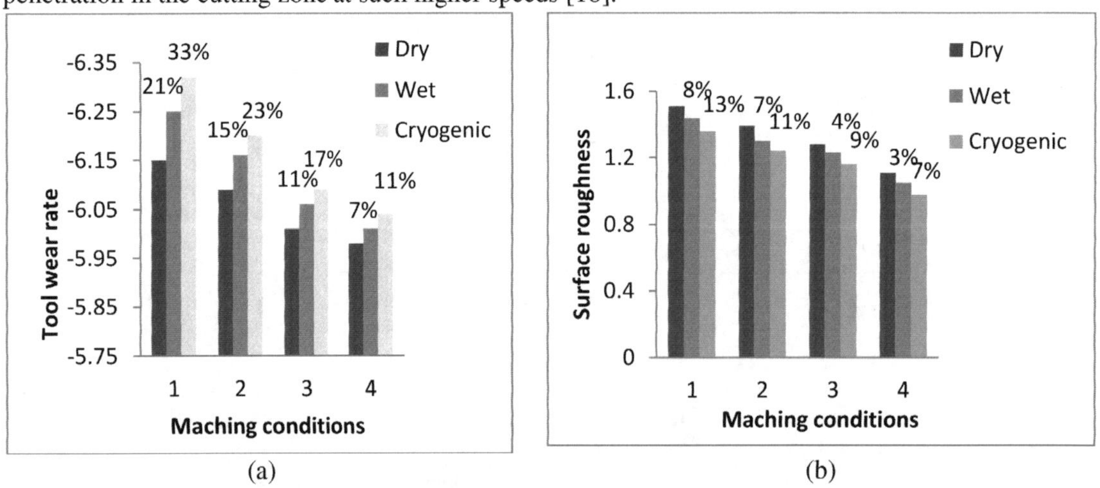

Figure 2. Comparison of cutting conditions (a) Tool wear rate (b) Surface roughness

3.2 Surface roughness analysis

Fig. 2 (b) compares surface roughness under varying cutting conditions. Surface roughness improves with increasing cutting speed because of the thermal softening of work piece. Both cryogenic and wet conditions produced better results than dry machining owing to the lubrication effect at the sliding surfaces interface [19]. As discussed earlier the benefits of cooling media is more pronounced at lower cutting speeds.

4. Conclusion

Present research carried out turning of Ti-6Al-4V under dry, wet and cryogenic conditions. Tool wear and surface roughness were the focus of attention under varying machining parameters. Following conclusions were drawn:

- Tool wear and surface roughness are highly dependent on cutting speed as the temperature increases with increasing speed.
- Cooling media has significantly improved tool life by swift extrication of heat produced and reducing cutting zone temperature. Use of cryogenic media and cutting fluid improved tool life by 33% and 21% respectively at lower speeds.
- Surface roughness was reduced due to the lubrication effect of cooling media. Surface finish improved by 13% and 8% using cryogenic and cutting fluid respectively. Improvements were more pronounced at lower cutting speeds.

Employment of cooling system has considerably improved productivity and economy of manufacturing system. However sustainability of the manufacturing system pose new challenges which needs deliberation. Future research includes the study of effects of varying feed rate and depth of cut in terms of productivity, economy and sustainability of manufacturing systems.

References

[1] A. Shokrani, V. Dhokia, and S. T. Newman, "Environmentally conscious machining of difficult-to-machine materials with regard to cutting fluids," *International Journal of machine Tools and manufacture,* vol. 57, pp. 83-101, 2012.

[2] M. Mia, A. Rifat, M. F. Tanvir, M. K. Gupta, M. J. Hossain, and A. Goswami, "Multi-objective optimization of chip-tool interaction parameters using Grey-Taguchi method in MQL-assisted turning," *Measurement,* vol. 129, pp. 156-166, 2018.

[3] P. Sivaiah and D. Chakradhar, "Effect of cryogenic coolant on turning performance characteristics during machining of 17-4 PH stainless steel: A comparison with MQL, wet, dry machining," *CIRP Journal of Manufacturing Science and Technology,* vol. 21, pp. 86-96, 2018.

[4] I. Jawahir, H. Attia, D. Biermann, J. Duflou, F. Klocke, D. Meyer, *et al.*, "Cryogenic manufacturing processes," *CIRP annals,* vol. 65, pp. 713-736, 2016.

[5] S. Yi, Y.-C. Jang, and A. K. An, "Potential for energy recovery and greenhouse gas reduction through waste-to-energy technologies," *Journal of Cleaner Production,* vol. 176, pp. 503-511, 2018/03/01/ 2018.

[6] L. N. López de lacalle, J. Pérez, J. I. Llorente, and J. A. Sánchez, "Advanced cutting conditions for the milling of aeronautical alloys," *Journal of Materials Processing Technology,* vol. 100, pp. 1-11, 2000/04/03/ 2000.

[7] M. C. Shaw and J. Cookson, *Metal cutting principles* vol. 2: Oxford university press New York, 2005.

[8] J. Barry, G. Byrne, and D. Lennon, "Observations on chip formation and acoustic emission in machining Ti–6Al–4V alloy," *International Journal of Machine Tools and Manufacture,* vol. 41, pp. 1055-1070, 2001/05/01/ 2001.

[9] B. D. Jerold and M. P. Kumar, "The influence of cryogenic coolants in machining of Ti–6Al–4V," *Journal of Manufacturing Science and Engineering,* vol. 135, p. 031005, 2013.

[10] S. Sun, M. Brandt, S. Palanisamy, and M. S. Dargusch, "Effect of cryogenic compressed air on the evolution of cutting force and tool wear during machining of Ti–6Al–4V alloy," *Journal of Materials Processing Technology,* vol. 221, pp. 243-254, 2015.

[11] I. Lee, V. Bajpai, S. Moon, J. Byun, Y. Lee, and H. W. Park, "Tool life improvement in cryogenic cooled milling of the preheated Ti–6Al–4V," *The International Journal of Advanced Manufacturing Technology,* vol. 79, pp. 665-673, 2015.

[12] M. J. Bermingham, S. Palanisamy, D. Kent, and M. S. Dargusch, "A comparison of cryogenic and high pressure emulsion cooling technologies on tool life and chip morphology in Ti–6Al–4V cutting," *Journal of Materials Processing Technology,* vol. 212, pp. 752-765, 2012.

[13] M. Mia, M. K. Gupta, J. A. Lozano, D. Carou, D. Y. Pimenov, G. Królczyk*, et al.*, "Multi-objective optimization and life cycle assessment of eco-friendly cryogenic N2 assisted turning of Ti-6Al-4V," *Journal of Cleaner Production,* vol. 210, pp. 121-133, 2019.

[14] M. J. Bermingham, J. Kirsch, S. Sun, S. Palanisamy, and M. S. Dargusch, "New observations on tool life, cutting forces and chip morphology in cryogenic machining Ti-6Al-4V," *International Journal of Machine Tools and Manufacture,* vol. 51, pp. 500-511, 2011.

[15] S. I. Jaffery and P. T. Mativenga, "Assessment of the machinability of Ti-6Al-4V alloy using the wear map approach," *The International Journal of Advanced Manufacturing Technology,* vol. 40, pp. 687-696, 2008.

[16] M. J. Bermingham, S. Palanisamy, and M. S. Dargusch, "Understanding the tool wear mechanism during thermally assisted machining Ti-6Al-4V," *International Journal of Machine Tools and Manufacture,* vol. 62, pp. 76-87, 2012.

[17] Y. Fan, Z. Hao, M. Zheng, and S. Yang, "Wear characteristics of cemented carbide tool in dry-machining Ti-6Al-4V," *Machining Science and Technology,* vol. 20, pp. 249-261, 2016.

[18] K. A. Venugopal, S. Paul, and A. B. Chattopadhyay, "Tool wear in cryogenic turning of Ti-6Al-4V alloy," *Cryogenics,* vol. 47, pp. 12-18, 2007.

[19] N. Dhar and M. Kamruzzaman, "Cutting temperature, tool wear, surface roughness and dimensional deviation in turning AISI-4037 steel under cryogenic condition," *International Journal of Machine Tools and Manufacture,* vol. 47, pp. 754-759, 2007.

Tall buildings seismic behavior comparative study by increasing the concrete mechanical strength through non-linear static analysis and seismic performance

F Lopez-De-La-Cruz[1], A Choncen-Arboleda[1], J Moreno-Sanchez[1,*]

[1] Ingeniería Civil, Universidad Peruana de Ciencias Aplicadas (UPC), Lima 15023, Perú.

*pccijmor@upc.edu.pe

Abstract. The use of high strength concrete in tall buildings implies less lateral displacement, however, an important question is. In what other measures the seismic behavior of a building improves with high strength concrete (550 kg/cm^2). As a result, the material allows the construction of slender structural elements, however we won't be able to know how much the ductility of the structure in front of a seismic event will improve. Therefore, the reason of this study is to analyze the dynamic response of a tall building with different sections and resistance to the compression of its structural elements. The aim of this study is to obtain the curve of capacity and the point of seismic performance of each structural model.

1. Introduction

During the last decade, earthquakes have caused economic loss as well as in terms of social costs. Since the year 2000, more than 1,000,000 people have died due to earthquakes, for which the associated cost was 445,700,000 million dollars [1]. Evidence of this was the Tangshan earthquake (China) that occurred in 1975 that killed 200,000 people. Also, the 1994 earthquake in Northridge (USA) left 75 dead and approximately 8,700 wounded. In turn, it had a loss of about 42 billion dollars. Likewise, the 1995 earthquake in Kobe (Japan) caused about 6,000 deaths and more than 120 billion in economic losses. On the other hand, in August 1996, the earthquake that occurred Izmet (Turkey) killed 20,000 people and caused economic losses of 12 billion. Finally, the Chi-chi (Taiwan) earthquake in 1999 caused an estimate of 8 billion loss, as did the Gujarat (India) earthquake in 2006 where there were around 18,000 dead and 330,000 demolished buildings [2].

High-strength concrete (HPC) has allowed the continuous development of high-rise buildings, with the purpose of using floor areas more efficiently and obtaining a better lateral response [3]. However, unknowns arise about the seismic behavior of the structure with the use of high strength concrete; particularly, in the plastic hinge mechanism, the structure ductility and its seismic performance [4]. It is also an advanced cement material with excellent durability, which offers a potential to become a practical solution to improve the sustainability of buildings and other infrastructure [5].

That's why, it is evidenced that with this concrete there are advantages such as the inertia moment decreases in the structural elements and as it had been mentioned a better durability, since it increases the structure life period in comparison with the normal resistance concrete.
However, for most countries, demonstration projects that used ultra-high strength concrete (UHPC) didn't generate acceptance. This is due to the lack of design codes, limited knowledge of both material, production technology and high costs, limiting the implementation of this material [6].

To address the problem of the structure seismic behavior by increasing mechanical resistance in Latin America, a non-linear static Push-over analysis is proposed for five models over 18 levels, increasing the strength of concrete in structural elements (plates , columns and beams) in order to obtain the capacity curve and the seismic performance point of the structure. For this, Model A has a resistance of 210 kg / cm2, Model B of 350 kg / cm2, Model C of 550 kg / cm2, Model D of 210-350 kg / cm2 and finally Model E of 350-550 kg / cm2.

In the buildings analyzed, the resulting structure implies that for a 60-meter building with compressive strengths of the E model (350-550 kg / cm2) was obtained a basal shear of 1349.80 tons, a maximum displacement of 48.75 cm and a maximum ductility of R = 6.

2. Method

This research evaluates the non-linear response of structural models taking into account the variation of the mechanical strength of concrete, with the aim of obtaining structures with greater resistance capacity, a reduction of the own weight and the decrease of its cost in the face of a lower volume of flexo-compression and cut, so it will have to be verified with a non-linear analysis and seismic performance.

The displacement that is generated in the upper level and the performance point of the building is done before a pattern of incremental lateral loads and constant gravitational loads; therefore, ductility, the mechanism of formation of plastic kneecaps and damage to different levels of seismic demand will be evaluated. The latter implies the progressive degradation of stiffness and the collapse of the structure.

The five models with different compressive strengths (210, 350, 550 kg / cm2 and two combinations of 210-350 and 350-550 kg / cm2) are analyzed by computer. They are defined as structural elements (beams, columns and plates) for which is proposed sections of (30x60) cm in main beams, (30x40) cm in secondary beams, (70x70) cm in central columns, (40x100) cm in non-central columns and plates with 50cm of thickness, all with a high-strength concrete of 550 kg / cm2.

On the other hand, the structures are considered to be embedded in the base, column nodes and walls. It should be noted that the slab was modeled as a “membrane” type element so that the loads are transmitted in their entirety and not lost due to the rigidity contribution.

For loads of the entire structure, national standards will be considered in the models. A spectral modal analysis is performed to evaluate these models, granting 100% of the dead load plus a percentage of live load and live roof load, depending on the type of building and its comparison with the maximum allowable drift of 0.007 [7].

That is why the use of high strength concrete in flexural elements exhibits greater rotational ductility, improves the capacity of the column, as well as increasing the stiffness of the "beam-column" joints [8]. For this investigation we want to analyze the seismic behavior with respect to the increase in concrete resistance. For this, in each structural model the sections of the structural elements are reduced, since the f'c is increased in order to know the improvement of the non-linear response to the concrete variation.

Subsequently, the design [9] of each element for each model was carried out in order to obtain a resistant earthquake structure and to be able to insert the plastic labels manually and automatically to the face of each structural element, since if you do not have the design of the entire building could not know the mechanism of failure of the structure.

Therefore, once the bending and cutting labels have been assigned to each element, the moment of inertia of the shear and bending moment is reduced by factors of 0.4 - 0.3 respectively for beams, 0.4 - 0.7 in columns and factors of 0.4-0.5 in walls structural causing the structure to lose stiffness due to the cracking factor introduced.

With this procedure, the capacity curve is obtained and the performance point is generated by the capacity spectrum method [10] where the capacity curve is converted into capacity spectrum to obtain pseudo-accelerations (Sa) and pseudo-displacements (Sd) for each point of the shear and its displacement as shown in (1), (2), (3) and (4):

IOP Conf. Series: Materials Science and Engineering **689** (2019) 012007 doi:10.1088/1757-899X/689/1/012007

$$S_a = \frac{V/W}{\propto_1} \tag{1}$$

$$S_d = \frac{\Delta_{tope}}{PF_1 \emptyset_{tope,1}} \tag{2}$$

$$PF_1 = \left[\frac{\sum_{i=1}^{N}(w_i\phi_{i1})/g}{\sum_{i=1}^{N}(w_i\phi_{i1}^2)/g}\right] \tag{3}$$

$$\alpha_1 = \frac{\left[\sum_{i=1}^{N}(w_i\phi_{i1})/g\right]^2}{\left[\sum_{i=1}^{N} w_i/g\right]\left[\sum_{i=1}^{N}(w_i\phi_{i1}^2)/g\right]} \tag{4}$$

Where:
v: shear force
w: structure weight
$\propto_1$: modal mass coefficient for mode 1
PF_1: seismic participation factor mode 1
wi/g: assigned mass of level i
$\emptyset_{i,1}$: amplitude of mode 1 at level i
Δ_{tope}: roof shift
N: N level, higher structure

Likewise, the seismic demand spectrum (Sa vs T) is converted to an ADRS format (Sa vs Sd), as shown in (5) and (6):

$$S_d = \frac{1}{4\pi^2} S_a T^2 \tag{5}$$

$$T = 2\pi \sqrt{\frac{S_d}{S_a}} \tag{6}$$

Where:
Sd: pseudo-displacement
T: period
S_a: pseudo-acceleration

Finally, after obtaining the performance point, the scale factor is iterated according to its return period. Therefore, for the present investigation a scale factor of 1.5 corresponding to a Tr = 1500 years is used in order to find the point of transfer. Thence, the maximum ductility of the structure resulting from the division between the last displacement and the point of yield is obtained.

3. Results

To confirm our method of analysis we chose a seismic and important capital of Latin America, we analyzed five models of a 20-story building with a rigid core. Also, this research was analyzed using

IOP Conf. Series: Materials Science and Engineering **689** (2019) 012007 doi:10.1088/1757-899X/689/1/012007

the ETABS 2016 version 16.2.0 software for the structural analysis and resistant earthquake of the five models.

3.1. Models's Structural Weight

Next, the results will be presented in Table 1, the weights found in the five models analyzed.

Table 1. Five model's weight

Models	Total Weight	f'c	Us
A	30707.63	210	tonf
B	29290.22	350	tonf
C	28727.99	550	tonf
D	29271.56	210-350	tonf
E	27458.88	350-550	tonf

Once the weights of the models have been found, a comparison and differences can be made to know when it varies in percentage, as presented in Table 2 below.

Table 2. Weights comparison and percentage

f'c-f'c		Dif.	U	%
210	350	1415.41	tnf	5
210	550	1979.64	tnf	6
350	550	564.23	tnf	2
210	210-350	1436.07	tnf	5
210	350-550	3248.74	tnf	11
350	210-350	20.67	tnf	0.07
350	350-550	1833.33	tnf	6
550	210-350	543.56	tnf	2
550	350-550	1269.1	tnf	4

3.2. Models Periods

The results of the periods of the five models are presented in Table 3.

Table 3. **Fundamental periods of the models analyzed in the xx direction**

Models	f'c	Period	U
A	210	1.969	s
B	350	1.903	s
C	550	1.698	s
D	210-350	1.856	s
E	350-550	1.71	s

3.3. Models's Dynamic Shear

Regarding the dynamic shear, they were found for both the X and Y shears of the five models. Which are presented in Table 4.

IOP Conf. Series: Materials Science and Engineering **689** (2019) 012007 doi:10.1088/1757-899X/689/1/012007

Table 4. Five models's dynamic cutters in X and Y

f'c	V(x)	V(y)	U
210	1525.8482	1726.2288	tonf
350	1458.8161	1650.8277	tonf
550	1384.9101	1626.1457	tonf
210-350	1642.9101	1622.2411	tonf
350-550	1349.9831	1575.3812	tonf

3.4. Maximum model drifts

The maximum allowable drifts were obtained for each structural model with its respective f'c as shown in Figure 1.

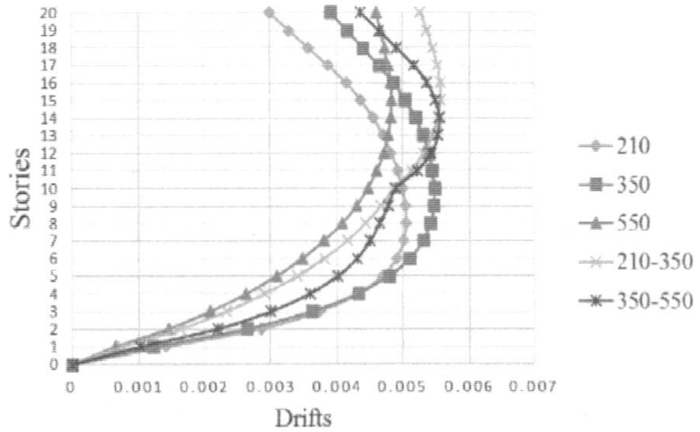

Figure 1. Maximum drift figure in "X".

Likewise, the displacement generated at the upper level and at the performance point of the building before a pattern of incremental lateral loads and constant gravitational loads; whence, ductility, the mechanism of plastics kneecaps formation and damage to different seismic's levels demand will be evaluated. The latter implies the stiffness's progressive degradation and the structure collapse as can be seen in the following Figure 2.

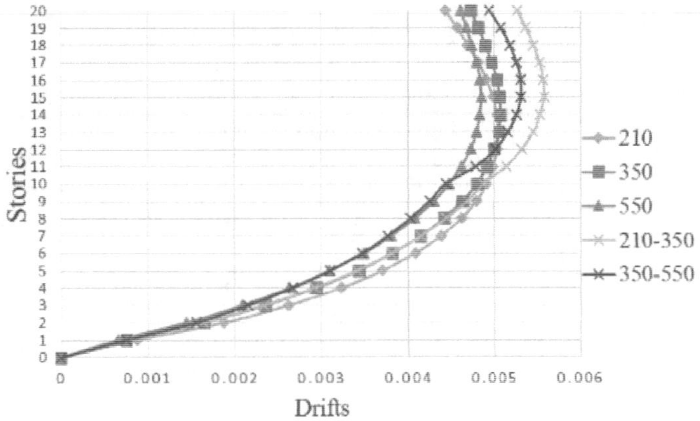

Figure 2. Maximum drift figure in "Y".

3.5. Models Capacity Curve

About the capacity curve, it was obtained in both X and Y, which can be seen in the following figures 3 and 4.

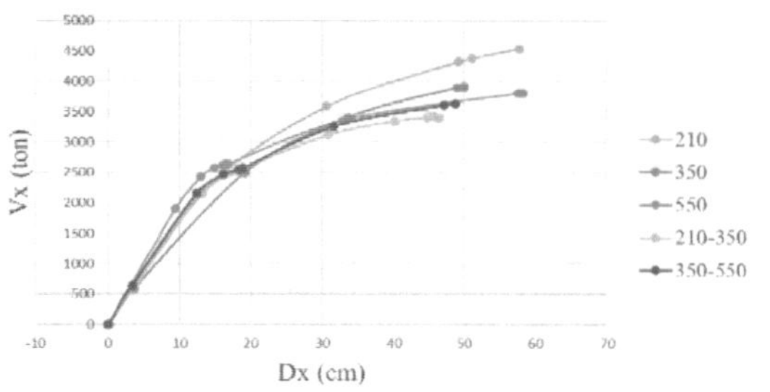

Figure 3. Capacity curve in "X".

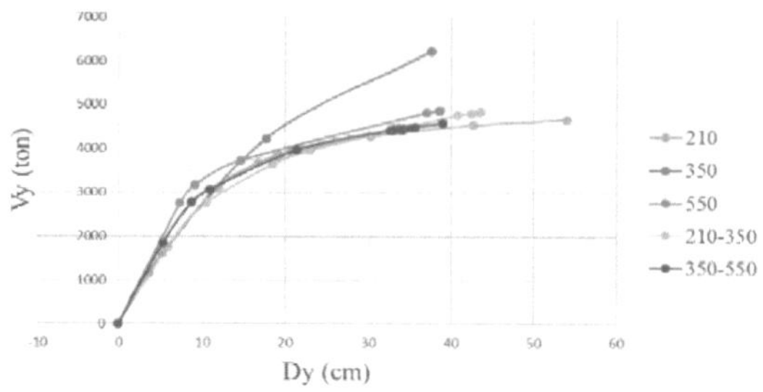

Figure 4. Capacity curve in "Y".

The results obtained of the weight structure, was very advantageous for model E, since it weighs 11% less compared to model A. This is because sections of less moment of inertia were used, due to the Increase in the modulus of concrete elasticity.

Also, in table 5 it is possible to show the total weight, the concrete volume used and the amount of formwork required for each model. Whereby, model E contemplates a 14.3% savings in concrete and 6.3% savings in formwork with respect to model A. However, for the remaining models the savings in concrete and formwork are within a range 3.7% -10.5%; 2.0% -4.6% respectively.

Table 5. Concrete and formwork quantity table regarding to model A

Models	Weight		Concrete		Formwork	
	Ton	Saving	M3	Saving	M2	Saving
A	30707.63	0.0%	10088	0.0%	51438.6	0.0%
B	29292.22	-4.6%	9663	-4.2%	50305.8	-2.2%
C	28727.99	-6.4%	9029	-10.5%	49054.6	-4.6%
D	29271.56	-4.7%	9710	-3.7%	50397	-2.0%
E	27458.56	-10.6%	8641	-14.3%	48203.8	-6.3%

On the other hand, table 6 shows the total cost of concrete per m3 and the total formwork cost per m2, for each model with its respective f'c. Therefore, model E has an increase of 2.2% of the total cost with respect to model A, while for the other models the percentage of increase in total cost is within a range of 3.5% - 9.6% with respect to the first model.

Table 6. Concrete and formwork total cost table regarding to model A

Models	f'c	Concrete		Formwork		Total	Saving
	Kg/cm2	M3	Cost($)	M2	Cost($)	Cost($)	%
A	210	10088	63.33	51438.56	12.62	1287880.70	0.0%
B	350	9663	79.70	50305.76	12.62	1404856.06	9.1%
C	550	9029	83.94	49054.56	12.62	1376822.65	6.9%
D	210	4740	63.33	24946.48	12.62	1332158.84	3.4%
	350	4970	79.70	25450.48	12.62		
E	350	4077	79.70	23873.28	12.62	1316148.85	2.2%
	550	4563	83.94	24330.48	12.62		

The periods of each structure were decreasing as the strength of the concrete increased, obtaining a shorter period in models C and E, having high strength concrete. For them, a more rigid structure was obtained limiting it to excessive displacements.

The results of the capacity curve that each model yielded according to the procedure of ASCE41-13 and FEMA 440 standards were the following.

- ✓ Model A: Maximum shear of 4536.81 ton, maximum displacement of 57.73 cm.
- ✓ Model B: Maximum shear of 3907.05 ton, maximum displacement of 50.10 cm.
- ✓ Model C: Maximum shear of 3810.46 ton, maximum displacement of 58.24 cm.

On the other hand, for models D and E, shear of 3399.46 and 3623.36 tons and maximum displacements of 46.40 cm and 48.75 cm respectively were obtained. However, the real ductility that was obtained in each model resulting from the division of the ultimate displacement and yield, as well as the seismic reduction force coefficient were the following:

- ✓ Model A: μ= 4.14 → R= 5
- ✓ Model B: μ= 2.54 → R= 3
- ✓ Model C: μ= 4.27 → R= 5
- ✓ Model D: μ= 3.91 → R= 5
- ✓ Model E: μ= 5.15 → R= 6

That is why, with these results it is shown that the E model is so far one of the most optimal models because it has greater ductility compared to the other models. Regarding the load patterns for each model, it was observed that for the E model, lower lateral force coefficients are obtained in both directions due to the combination of the modulus of elasticity that the structure possesses.

Finally, regarding the seismic performance points that were found based on the capacity curve and different demand spectra, they were the following: With period and effective damping of Teff and Beff.

- ✓ Model A: Teff = 2.506, Beff= 0.112 and a performance of D= 26.486 cm. was obtained.
- ✓ Model B: Teff = 3.034, Beff= 0.1844 and a performance of D= 44.71 cm. was obtained.
- ✓ Model C: Teff= 2.161, Beff= 0.1221 and D= 18.87 cm.
- ✓ Model D: Teff= 2.396, Beff= 0.1296 and D= 21.01 cm.
- ✓ Model E: Teff= 1.672, Beff= 0.1338, D= 14.46 cm.

4. Conclusions

As a final conclusion based on the results provided in the previous paragraphs, it was shown that model E, has a better seismic behavior with respect to the other models, because the structure is more rigid, it presents a reduction of its total weight in 11% with regard to the model A which has a normal resistance concrete, also, its ductility is 5.15 higher than all models and its performance point is within the ranges of life safety (LS), therefore, the structure is oversized, since the standards recommend that for this type of building (offices) the structure be within the immediate occupation (IO) range. On the other hand, in terms of the costs generated by the construction of a building with high-strength concrete, the E model represents a higher cost of 2.2% compared to the model with lower compression resistance, which is very satisfactory since a high strength concrete has a higher cost compared to a normal strength. However, it should be noted that the structure is in the life safety range, which indicates that the cross sections of the elements can be reduced by a greater percentage, leading to less use of concrete and formwork, thus reducing the structure total cost.

Finally, it is demonstrated that the use of high-strength concrete together with other less resistant concrete not only provides good seismic behavior and a much lower cost, but also, the reduction of construction materials provides several environmental benefits, which, makes it an extremely profitable project in the short and long term and especially that is in balance with the ecosystem.

References

[1] G. Risks "Earthqueakes and their socio-economic consequences", (https://link.springer.com/referenceworkentry/10.1007/978-3-642-35344-4_259), 2013.

[2] A. Nolan, "Handbook of Earthquake Engineering", 2015.

[3] A. Mwafy, "Assessment of Material Strengrh Implications on Seismic Design of Tall Buildings through Collapse Analysis", 2015.

[4] American Society of Civil Engineers, ASCE 41-13, "Seismic evaluatioHihn and retrofit of existing buildings", 2014.

[5] M. Schmidt, "High Performance Concrete": Research Developmente and Applocation in Europe. The 7th International Symposium on the utilization of High Strength7High Performance Concrete, ACI, Washington DC, USA, 2005: 51-78.

[6] YL. Voo, "Ultra High Performance Concrete"- Technology for Present and Future. ACI Singapore, Building Construction Authority Joint Seminar on Concrete for Sustainability, Productivity and The Future, 2017.

[7] Reglamento Nacional de Edificaciones, Norma Técnica Peruana de Edificación E.030 "Diseño Sismo Resistente", 2018.
[8] R. Pendyala, "Full Range Behavior of High-Strength Concrete Flexural Members, Part 1", ACI Structural Journal, V93, N1, 1996, --30-35..
[9] Reglamento Nacional de Edificaciones Norma Técnica Peruana de Edificación E.060 "Concreto Armado", 2009.
[10] FEMA 440, Departmentt of Homeland Security Federal Emergency Management Agency: "Improvement of nonlinear static seismic analysis procedures", 2005.

IOP Conf. Series: Materials Science and Engineering **689** (2019) 012008 doi:10.1088/1757-899X/689/1/012008

Analysis of surface treatment of ASTM A516 Grade 70 using Salt spray method

Muhammad Ali Khan[1,a], Aqueel Shah[2,b], Syed Husain Imran Jaffery[1,c], Mushtaq Khan[1,d], Adnan Rasheed Khan[3,e]

[1]Department of Design and Manufacturing Engineering (DME), School of Mechanical and Manufacturing Engineering (SMME), National University of Sciences and Technology (NUST), Sector H-12, 44000, Islamabad, Pakistan
[2]Pakistan Naval Engineering College (PNEC), National University of Sciences and Technology (NUST)PNS Jauhar, Habib Ibrahim Rehmatullah Road, Karachi – 75350, Pakistan
[3]Comsats university Islamabad (CUI)

Email: [a]muhammadali.pg@smme.edu.pk, [b]aqueel@pnec.nust.edu.pk, [c]imran@smme.edu.pk, [d]mkhan@smme.edu.pk, [e]ardkhan138@comsats.edu.pk

Abstract. Conversion coating is an effective way to safeguard against corrosion. It is performed on a number of materials inhibiting the surface layer to corrosive attack. In this work surface engineering is carried out on test material ASTM A 516 Grade 70 which is widely used for constructional purposes. During the passivation process samples were given a hot alkaline oxide bath in sodium hydroxide and sodium nitrate. The corrosion resistance was then analyzed using advanced testing procedures. Salt spray test was used as a yard stick in making a comparative analysis of different samples including unpassivated samples, passivated samples and passivated samples with prior surface treatment. Results have shown that mechanical surface treatment improves material life by 3% whereas passivation with prior surface treatment can increase life by 18%.

Keywords: Surface Engineering, Passivation, ASTM A516 Grade 70, Salt spray method,

1. Introduction

Metals, which makes up most of the periodic table (91 out of 118), are elements which posses good thermal and electrical conductivities[1]. They are opaque and shiny as well as ductile, malleable and fusible. Material especially metals undergo chemical reactions with the surrounding and are transformed into a more stable shapes. Corrosion is a combination of oxidation-reduction reaction collectively known as redox reactions[2]. The base metal acting as anode gets oxidized primarily by the oxygen present in the air. As a result metal ion is produced and free electrons are librated which reduces oxygen forming hydroxide ions in presence of water. Final product is Iron Oxide (Fe_2O_3) commonly known as rust. This is a highly unstable and readily disintegrating product which causes pitting and adversely affects mechanical strength. A numbers of factors govern the initiation, propagation and transformation of corrosion [3]. A few act as basic ingredients required for the process of corrosion while others acts as catalysts. The reactivity of metal known as oxidation state defines the speed of redox reactions[4]. Highest oxidation state is that of iridium tetraoxide cation, i.e +9 and lowest is -5 for boron, gallium, indium and thalium. Corrosion causes huge losses in different industries by wasting resources and degrading finished and unfinished materials. According to one estimate corrosion damage amounts to about 4% of Gross Domestic Product (GDP) of a country[5]. Researchers have used different methods to counter the process of corrosion. A detailed study was carried out by M. Fenker et al[6] on anti corrosive properties of hard coatings on steel surface. Film growth, microstructure and defects were analyzed. It was concluded that a 10 μm thick single layered nitrite coating is effective against corrosion. Another study [7] showed that surface treatment plays a vital role in the subsequent anti corrosion effectiveness of coatings. Surface treatment in some cases alters the mechanical structure of surface layer whereas thermal and electrical conductivities can also be altered which significantly affects the coating properties. Passivation process or conversation coating is a highly efficient process in which the outer surface of a corrodible material is made resistant to corrosion. H. Zhu et al [8] carried out a process called hot alkaline nitrate (black oxidizing) bath. This process deposits magnetite Fe_3O_4 on the outer surface which protects the base material from corrosion. E. Onofre-Bustamente et al[9] highlighted that conversion coating in addition to the corrosion protection also increases the adhesiveness of coating to substrate. E. Banczek[10] showed that corrosion resistance of carbon steel can be substantially increased by adding benzotrizole to zinc Phosphating bath. Research carried out by Y.-W. Choi et al [11] compared the effects of anodizing, annealing and combination of both in terms of corrosion protection. Corrosion protection by single annealing was in amorphous phase and comparatively weak. A positive shift in tafel plot was observed by corrosion potential when anodizing and annealing were used in combination. S. B. Abusuilik et al [12] carried out a research to examine the effects of intermediate surface treatments. These included micro blasting with two types of blasting media, Ar ion Plasma etching followed by Ti ion bombardment. It was shown that the use of intermediate surface treatment greatly enhances the performance of cathode arc coatings. Research motivation for this study is to explore the benefits of passivation in combination with surface treatment. The aim is to find an effective yet economical way to increase corrosion protection without compromising on strength.

2. Experimental setup

ASTM A516 grade 70 was selected for experimentation and analysis. It is largely used for constructional purposes where its high strength and durability is utilized. Its chemical composition is given in Table 1. Test material is also the preferred choice industrially making pressure vessels and boilers because of its toughness. Its mechanical properties are given in Table 2.

Table 1. Chemical Composition (Wt %) of ASTM A516 grade 70

C = 0.10 – 0.28%	Mn = 0.85 – 1.2%	P = 0.010 – 0.035%	S = 0.010 – 0.035%
Si = 0.15-0.40%			

Table 2. Mechanical properties of ASTM A516 grade 70

Yield Strength	424 N/mm^2
Tensile Strength	511 N/mm^2
Elongation	24 %

3. Design of experiment – Sample preparation

Different types of samples were prepared as tabulated in Table 3 with dimensions of 8 x 5 inches. Type 1 was untreated samples cut from the original material. Type 2 samples were made by surface treating the original samples. An emery paper grinder was used for surface treatment. RPM was set at 3000 revs and grit size of 300 μm was used. Type 3 samples were made by treating original material in hot alkaline oxide solution for 45 minutes at 130°C. The solution had the aqueous mixture of sodium Nitrate (NaNO3) and sodium Hydroxide (NaOH) in 1:3ratio. An outer layer of Fe_3O_4 (magnetite) is formed which prevents base material from corrosion. Type 4 was prepared by first surface treating the samples by emery paper grinding and then giving them the hot alkaline passivation bath. These sample types are shown in Fig. 1. Three samples of each type were made and tested for taking average values.

Table 3. Design of experiment – Different types of samples

Sample	Surface Treatment	Conversion Coating
Type 1	x	x
Type 2	√	x
Type 3	x	√
Type 4	√	√

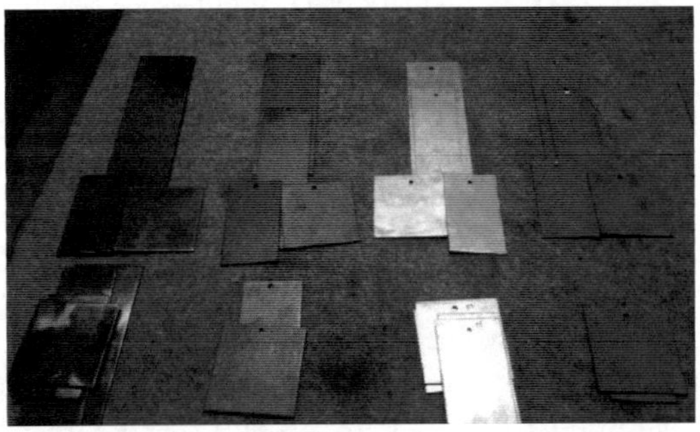

Figure 1. Prepared Samples Type 1 to Type 4 (R to L)

4. Experimentation

Samples were then kept in salt spray chamber as shown in Fig. 2. In this test the samples were exposed to an atomized 5% NaCl solution for 48 hours in a controlled environment at 40°C temperature and 60% humidity. Such atmosphere is strongly pro corrosive as it attacks the surface and accelerates the phenomenon of corrosion. After Salt spray test, bending test of samples was carried out on Tinius Olsen H150KU with bending testing kit installed as shown inFig. 3. The samples were bent under force up to and beyond their elastic limit until they bend plastically.

Figure 2. Salt Spray chamber

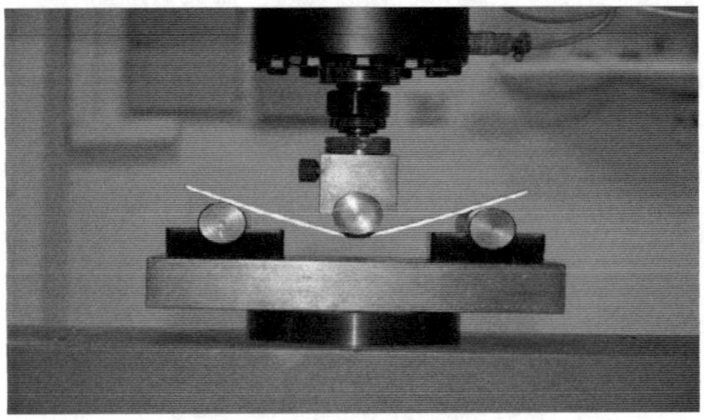

Figure 3. Bending test

5. Results and Analysis

The initial and final weights were calculated and compared. The weight loss and strength results are as shown in Table 4.

Table 4. Salt spray and bending test results

Sample Type	Salt Spray Results		Bending Test Results
	Initial Weight (g)	**Final Weight (g)**	**MPa**
1	232.82	209.63	729
2	237.97	230.86	782
3	233.74	222.48	761
4	234.50	233.44	791

5.1 Weight loss analysis
Weight loss is a direct indication of the extent of corrosion that has occurred. The corroded portion of the sample was washed away and the weight loss was analyzed as shown in Fig. 4. It was evident from the results that Type 4 sample was the least corroded because of surface treatment and passivation. Type 2 gave the second best result. Its good performance was because of the surface passivation which resisted corrosion. Type 3 was only surface treated and offered even lesser resistance. Type 1 had no protection against corrosion and lost significant amount of weight due to redox reactions.

5.2 Strength analysis
The purpose of this test was to analyze any strength reduction that has occurred to the original material because of the surface treatment or surface passivation process. The results displayed in Fig. 5 shows that strength has not been compromised because of the surface treatment procedures. Interestingly it was observed that strength was improved because of the surface treatment and passivation. The reason here is the change in surface morphology and it can well be concluded that the treatment for improving corrosion resistance has not compromised on strength.

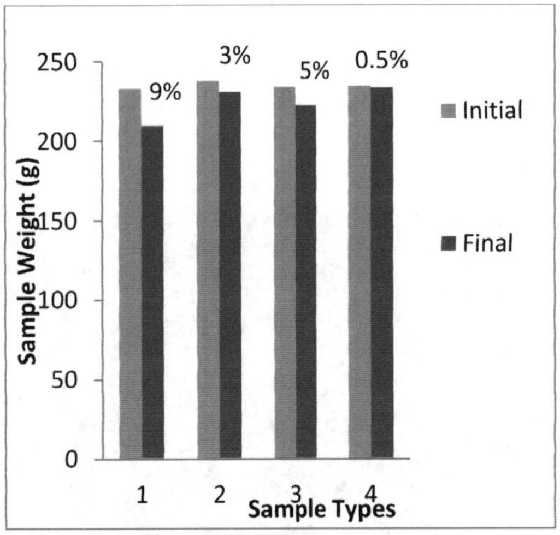

Figure 4. Weight loss analysis

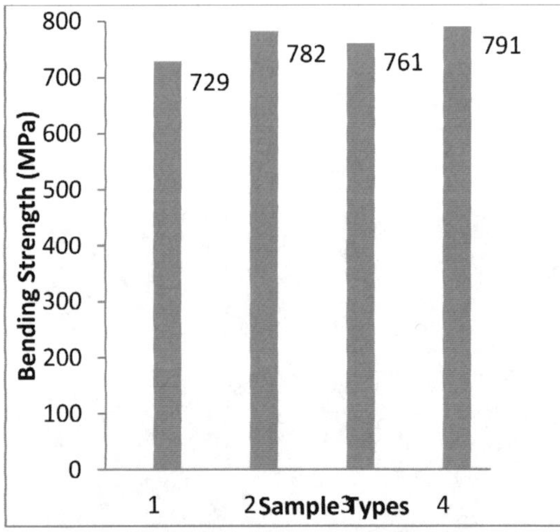

Figure 5. Strength analysis

6. Conclusion
During the course of experimentation, testing and analysis following conclusions were reached:-

1. ASTM A516 Grade 70 is a corrodible alloy of iron and needs to be protected as Type 1 sample corroded 9% of its body weight.
2. Surface treatment of test material by emery paper grinding has proved to be a simple yet effective way of resisting corrosion. It improved life by 3%.

3.Conversion coating on test material has produced good protection against corrosive attack. It passivates the outermost layer by oxidizing.
4.Surface treatment prior to conversion coating adds manifolds to the potential barrier of resistance to corrosion by increasing material life around 18%.
5.Strength is not compromised by surface treatment using emery paper grinding or conversion coating using hot alkaline oxide bath as evident by bending tests.

References

[1] B. N. Popov, *Corrosion engineering: principles and solved problems*: Elsevier, 2015.
[2] W. Gao, *Developments in high temperature corrosion and protection of materials*: Elsevier, 2008.
[3] Z. Ahmad, *Principles of corrosion engineering and corrosion control*: Elsevier, 2006.
[4] M. Görlin, P. Chernev, J. Ferreira de Araújo, T. Reier, S. r. Dresp, B. Paul, *et al.*, "Oxygen evolution reaction dynamics, faradaic charge efficiency, and the active metal redox states of Ni–Fe oxide water splitting electrocatalysts," *Journal of the American Chemical Society,* vol. 138, pp. 5603-5614, 2016.
[5] L. L. Shreir, *Corrosion: corrosion control*: Newnes, 2013.
[6] M. Fenker, M. Balzer, and H. Kappl, "Corrosion protection with hard coatings on steel: Past approaches and current research efforts," *Surface and Coatings Technology,* vol. 257, pp. 182-205, 2014.
[7] J. Carew and A. Abdullah, "The influence of heat treatment on the corrosion behaviour of ferritic stainless steel solar panels in hot municipal water," *Corrosion science,* vol. 34, pp. 217-231, 1993.
[8] H. Zhu, F. Cao, D. Zuo, L. Zhu, D. Jin, and K. Yao, "A new hydrothermal blackening technology for Fe3O4 coatings of carbon steel," *Applied Surface Science,* vol. 254, pp. 5905-5909, 2008.
[9] E. Onofre-Bustamente, M. A. Dominguez-Crespo, J. Genescá-Llongueras, and F. Rodríguez-Gómez, "Characteristics of blueing as an alternative chemical conversion treatment on carbon steel," *Surface and Coatings Technology,* vol. 201, pp. 4666-4676, 2007.
[10] E. Banczek, P. Rodrigues, and I. Costa, "Investigation on the effect of benzotriazole on the phosphating of carbon steel," *Surface and Coatings Technology,* vol. 201, pp. 3701-3708, 2006.
[11] Y.-W. Choi, S. Shin, D.-W. Park, and J. Choi, "Surface treatment of iron by electrochemical oxidation and subsequent annealing for the improvement of anti-corrosive properties," *Current Applied Physics,* vol. 14, pp. 641-648, 2014.
[12] S. B. Abusuilik and K. Inoue, "Effects of intermediate surface treatments on corrosion resistance of cathodic arc PVD hard coatings," *Surface and Coatings Technology,* vol. 237, pp. 421-428, 2013.

IOP Conf. Series: Materials Science and Engineering **689** (2019) 012009 doi:10.1088/1757-899X/689/1/012009

Effect of Internal BN Substitution on Electronic Properties of Pyrene Derivatives

Li An

International Department of Shude High School, 610000, Chengdu, China

Email: lianworkmail@foxmail.com

Abstract. Pyrene as an important chemical building block has found wide applications in the field of organic electronics. In solution, pyrene and its derivatives are well-known examples for excimer formation. Here, we studied the boron-nitrogen (BN)-substituted pyrene, decorated with thiophene side chain. From density-functional theory (DFT) calculation, this isoelectronic BN substitution maintains the coplanarity of the central motif and the delocalized π-electron distribution, but additionally induces a much stronger permanent dipole moment. Time-dependent DFT indicates a much-reduced oscillator strength for low energy transitions, and surprisingly a forbidden π-π* transition, therefore resulting in a much red-shifted absorption spectrum. Our simulation results call for future experiments to study the photo-excited properties of BN-substituted pyrene and molecular self-assembly through dipole-dipole interaction and π-π* interaction.

1. Introduction

The study of organic electronics has long been viewed as an essential determinant in the field of display technology, renewable energy and flexible electronics.[1] In the past 50 years, photochemists have been devoting continuous efforts in researching organic semi-conductors[2], and their integration into high-performance organic electronic devices. Pyrene is one of the simplest π-conjugated molecules which has shown electrical conductivity undergoing electrochemical doping.[3] This feature, together with pyrene's unique radiometric properties[4] and its concentration-dependent excimer[5] gives rise to the multiple applications including luminescence-switching chemical sensing[4] and membrane biophysics.[5]

Owing to its outstanding optical and electronic properties, pyrene has represented an important building block in optoelectronic devices, such as organic light-emitting diodes (OLEDs)[6], organic field effect transistors (OFETs)[7], and organic photovoltaics (OPVs).[8] In OLEDs, pyrene's high charge carrier mobility and prominent hole injection ability, together with its high photoluminescence efficiency, is particularly suitable for deep blue OLEDs[9] compared with other chromophores[10] A high carrier mobility is also assured to fabricate OFET.[10], [11] OPVs based on pyrene molecules have showed a decent power conversion efficiency partly due to its high fill factor benefiting from the strong molecular self-assembly.[3]

Excimer formation from traditional pyrene has its unique applications in biology.[12],[13] However, the excimer formation is not always advantageous in organic electronics, given that photoluminescence efficiency of pyrenes notably decreases in thin films.[10] BN substitution has then been introduced as a tool to modulate the electronic properties of conjugated molecules, and there are isoelectronic analogues like naphthalene[14], PAHs pyrene[15], and phenanthrene.[16], [17] One further note is that unlike the common central substitution cases, in certain cases of C/BN hybrid fullerene, the BN substitution takes place out of the central area.[18]

Here, we use a DFT method to simulate the electronic properties of pyrene and its BN-substituted derivatives with thiophene as the side chain. Frontier orbital electron distribution and their energy levels are calculated. Furthermore, we use time-dependent DFT to calculate the excited state properties such as the absorption spectrum. Finally, we discuss our results in the context of excimer formation.

2. Results and Discussion

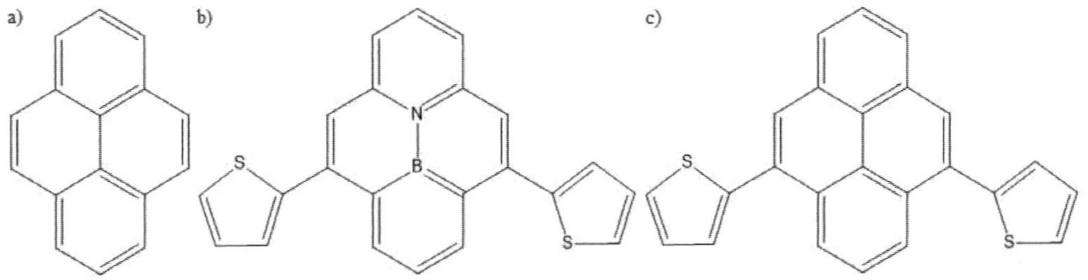

Figure 1. The skeletal structure of a) pyrene. b) BN substituted pyrene with thiophene sidechain. c) CC pyrene with thiophene sidechain.

The skeletal formula of pyrenes are shown in Figure 1 and rough Cartesian coordinates are then obtained before carrying out geometry optimization in Orca, an open-source DFT calculation software. Further notes concerning method is described in later sections. On exploring the electronic properties of BN pyrene (BNP) and CC pyrene (CCP), thiophene end group is covalently connected to pyrene to increase the solubility of the molecule for solution processing.

2.1 Geometry optimization

Though the pyrene center is coplanar, the thiophene substition is not in the same plane as the coplanar backbone in Figure 2, forming an angle to the central pyrene with approximately 120° on both sides. The substitued BN pyrene also formed an angle around 120° on both sides, but in this scenario the left and right C-C-S angles connecting pyrene and thiophene differ slightly less (see Table 1). Reason of this difference might be increased structural sustainbility after substitution, which to a certain extent reduces the induced dipole moments in thiophene's rotation around the centre when in motion. Since the planarity is related to the charge delocalization, and HOMO/LUMO, it could be deduced that thiophene sidechain will not significantly change pyrene's electronic properties, but mainly increases its solubility in solution. However, the side-chain has a significant conseqnece in the solid state through the molecular packing.

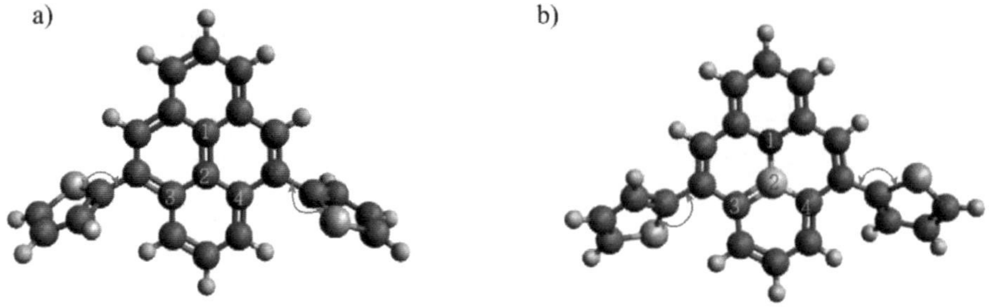

Figure 2. The molecular geometry visualization of a) CC pyrene. B) BN pyrene.

Table 1. The bond angles and bond lengths of CCP and BNP.[a]

Molecule	Left C-C-S angle (°)	Right C-C-S angle (°)	R_{12} (Å)	R_{23} (Å)	R_{34} (Å)	Total dipole moment (Debye)
CCP	120.5	121.7	1.435	1.420	1.419	0.300
BNP	120.9	122.1	1.468	1.511	1.511	2.694

[a]Note: Left and Right C-C-S angles are marked in Figure 2. R_{ij}, represents the distance between atom i and j labelled in Figure 2.

BN substitution was observed to change the bond length. As in the Table 1, the BN substitution increases the center C-C bond length, while also enlarging the difference between central bond and bonds connecting to the substitution area. Reason for such extension might due to change in the hybrid method (HF) caused by the substitution. The substitution also introduces a clear increase in total dipole moment, demonstrating a change in pyrene's polarity with BN replacing CC.

2.2 HOMO and LUMO analysis

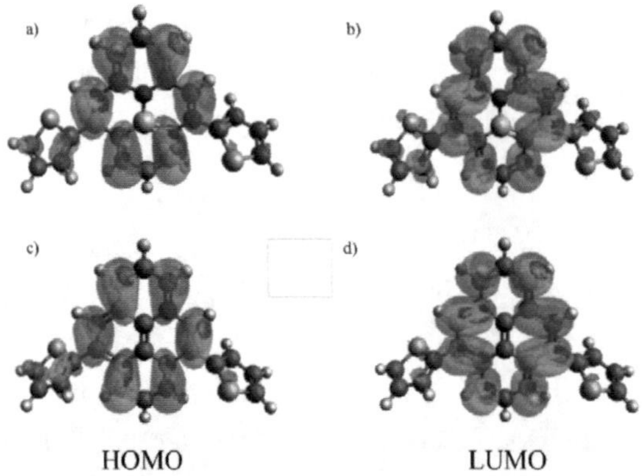

Figure 3. The frontier orbitals of pyrenes from DFT calculation. a) HOMO of BNP. b) LUMO of BNP. c) HOMO of CCP. d) LUMO of CCP.

Table 2. HOMO/LUMO specific data.

Molecule	HOMO-1 (eV)	HOMO (eV)	LUMO (eV)	LUMO+1 (eV)
BNP	-7.064	-7.036	1.217	1.501
CCP	-8.262	-7.268	1.541	2.443

This part is about the molecular orbital analysis of BN-CC pyrenes with thiophene groups. The frontier orbital distributes in an approximately symmetrical way relative to the center substitution area, where there almost none electron distribution exists. Band gaps (see Table 2) of 8.253eV (or equivalently, 150.2nm) and 8.809ev (140.7nm) are observed respectively in BNP and CCP. The abridged band gap value is largely due to the decrease in LUMO, which gives the CCP a stronger tendency to deduct electricity and a more violet-shift nature to the end of absorption.

2.3 Molecular absorption spectrum

It is worth noting that redshift occur when undergoing the BN substitution, the main highest absorption peak in Figure 4 which is locating around 5.5 eV, has moved leftward by 0.1 eV. The red shift indicates a decrease in BN band gap, a result perfect consistent with previous works done in the field with graphdiyne.[19] Moreover, the substitution introduces a new lower peak absorption situated at 4.95eV with an intensity peak of 0.571, despite the approximation attempt of set Gaussian width.

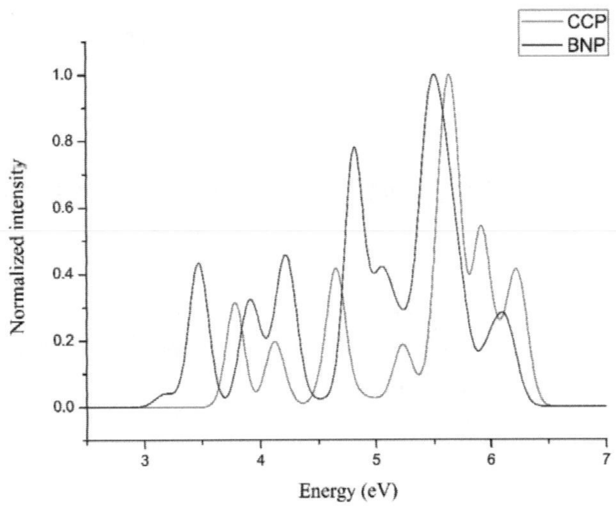

Figure 4. Normalized absorption spectrum of a) BNP-T, Gaussian width set as 0.1. b) CCP-T, Gaussian width set as 0.09.

Table 3. Transition dipole moment at the transitions with the lowest energies. a) CCP. b)BNP

Molecule	Wavelength (nm)	Intensity
	330.20	0.053
CCP	327.30	0.212

	303.30	0.030
BNP	428.50	0.000
	390.00	0.031
	357.50	0.346

The transition dipole in Table 3 moment results surprisingly reveal a forbidden π-π^* transition in substituted pyrene, as the data extracted shows that at a wavelength of 428.50 nm where HOMO to LUMO transition occur, the intensity is zero. The transition from HOMO-1 to LUMO and HOMO to LUMO+1 also just weakly allowed with 0.031 and 0.346 intensity respectively.

3. Method

The numerical calculation is performed using Orca 4.0.[20],[21] Orca, developed by Max Planck Institute, could conduct ab initio, DFT, and semi-empirical molecular orbital calculation. For the geometrical optimisation, the basic set is def2-TZVP. B3LYP. In TDDFT calculation, the Coulomb and exchange integral are approximated by RIJCOSX in Orca.

Calculated results are visualized via Avogadro for molecular orbitals and Gabedit[22] for UV-Vis spectrum.

3.1 Input Block

In this section, we use RHF, B3LYP and TightSCF as the main input blocks for calculating pyrene's electronic molecular properties.

3.2 Basis Set

The basis sets applied in the simulation mostly diffuse from the Rappoport property-optimized def2 basis set. The def2-SVP basis is used in geometry optimization of pyrenes as to split valence polarization, on which the new polarization functions modify the expression of the zeta function's expression on the def2 basis set. In the later part of excited states geometry optimization, def2/J is exercised with the Weigend's universal Coulomb fitting. The presence of ECPs beyond Kr is assumed. Despite longer time needed (see figure 4), an updated triple-zeta polarization function diffused from def2 basis set is applied to the PES scan, as the TZ polarization function better suits intercorrelation between H, S, C.

4. Conclusion

To sum up the research, we numerically compared the electronic properties of pyrene and BN-substituted derivative. Due to the isoelectronic nature, the coplanarity and the pi-conjugation is maintained, despite prolonged substituted area bond length. After substitution, the bandgap is smaller from the red-shifted onset of the UV-Vis spectrum, with changes occur in the electron clouds distribution. Surprisingly, we find the lowest energy transition is forbidden, probably due to symmetrical constrains. The introduced permanent dipole moment might significantly change the intermolecular interaction. Our single molecule calculation indicates that the excimer formation, traditionally popular in pyrene molecules, might change in the BN pyrene. Future experimental studies should focus on studying its photo-excited properties and molecular self-assembly in the solid state phase.

References

[1] Shaw, J. M. & Seidler, P. F. Organic electronics: Introduction. IBM J. Res. Dev. 45, 3–9 (2001).

[2] Forrest, S. R. & Thompson, M. E. Introduction: Organic electronics and optoelectronics. *Chemical Reviews* **107**, 923–925 (2007).

[3] Facchetti, A. π-Conjugated polymers for organic electronics and photovoltaic cell applications. *Chem. Mater.* **23**, 733–758 (2011).

[4] Karuppannan, S. & Chambron, J. C. Supramolecular chemical sensors based on pyrene monomer-excimer dual luminescence. *Chem. - An Asian J.* **6**, 964–984 (2011).

[5] Somerharju, P. Pyrene-labeled lipids as tools in membrane biophysics and cell biology. *Chem. Phys. Lipids* **116**, 57–74 (2002).

[6] Kuroda, H., Yoshihara, K. & Akamatu, H. The Spectral Dependence of the Photoconduction and the Mobility of the Charge Carrier in a Single Crystal of the Pyrene-Tetracyanoethylene Complex. *Bull. Chem. Soc. Jpn.* **36**, 1365–1366 (1963).

[7] Yamaguchi, S., Akiyama, S. & Tamao, K. Tri-9-anthrylborane and Its Derivatives: New Boron-Containing π-Electron Systems with Divergently Extended π-Conjugation through Boron. *J. Am. Chem. Soc.* **122**, 6335–6336 (2000).

[8] Yamaguchi, S. & Tamao, K. A Key Role of Orbital Interaction in the Main Group Element-containing π-Electron Systems. *Chem. Lett.* **34**, 2–7 (2004).

[9] Chercka, D., Yoo, S. J., Baumgarten, M., Kim, J. J. & Müllen, K. Pyrene based materials for exceptionally deep blue OLEDs. *J. Mater. Chem. C* **2**, 9083–9086 (2014).

[10] Figueira-Duarte, T. M. & Müllen, K. Pyrene-based materials for organic electronics. *Chemical Reviews* **111**, 7260–7314 (2011).

[11] Wang, Y. *et al.* 1-Imino nitroxide pyrene for high performance organic field-effect transistors with low operating voltage. *J. Am. Chem. Soc.* **128**, 13058–13059 (2006).

[12] Nishizawa, S., Kato, Y. & Teramae, N. Fluorescence Sensing of Anions via Intramolecular Excimer Formation in a Pyrophosphate-Induced Self-Assembly of a Pyrene-Functionalized Guanidinium Receptor. *J. Am. Chem. Soc.* **121**, 9463–9464 (1999).

[13] 1. Andriessen, R., Boens, N., Ameloot, M. & De Schryver, F. C. Non-a-priori analysis of fluorescence decay surfaces of excited-state processes. 2. Intermolecular excimer formation of pyrene. *J. Phys. Chem.* **95**, 2047–2058 (1991).

[14] Paetzold, P., Stanescu, C., Stubenrauch, J. R., Bienmüller, M. & Englert, U. 1-Azonia-2-boratanaphthalenes. *Zeitschrift fur Anorg. und Allg. Chemie* **630**, 2632–2640 (2004).

[15] 1. Dewar, M. J. S. & Dietz, R. 546. New heteroaromatic compounds. Part III. 2,1-Borazaro-naphthalene (1,2-dihydro-1-aza-2-boranaphthalene). *J. Chem. Soc.* 2728 (1959). doi:10.1039/jr9590002728

[16] 1. Dewar, M. J. S., Kubba, V. P. & Pettit, R. 624. New heteroaromatic compounds. Part I. 9-Aza-10-boraphenanthrene. *J. Chem. Soc.* 3073–3076 (1958). doi:10.1039/JR9580003073

[17] Jaska, C. A. *et al.* Triphenylene analogues with B2N2C2 cores: Synthesis, structure, redox behavior, and photophysical properties. *J. Am. Chem. Soc.* **128**, 10885–10896 (2006).

[18] Pattanayak, J., Kar, T. & Scheiner, S. Boron-nitrogen (BN) substitution patterns in C/BN hybrid fullerenes: C60-2x(BN)x (x = 1-7). *J. Phys. Chem. A* **105**, 8376–8384 (2001).

[19] Bu, H. *et al.* Isoelectronic doping of graphdiyne with boron and nitrogen: Stable configurations and band gap modification. *J. Phys. Chem. A* **116**, 3934–3939 (2012).

[20] Neese, F. The ORCA program system. *Wiley Interdiscip. Rev. Comput. Mol. Sci.* **2**, 73–78 (2012).

[21] Neese, F. Software update: the ORCA program system, version 4.0. *Wiley Interdiscip. Rev. Comput. Mol. Sci.* **8**, e1327 (2018).

[22] Allouche, A.-R. Gabedit—A graphical user interface for computational chemistry softwares. *J. Comput. Chem.* **32**, 174–182 (2011).

Alpaca Waste Management Model for Improving Fiber Productivity Through the use of Biomass in Andean Highland Communities

V Fernandez-Lizama[1], A Espinoza-Ccatamayo[1], P Chavez-Soriano[1], C Raymundo-Ibañez[2]

[1] Ingeniería Industrial, Universidad Peruana de Ciencias Aplicadas (UPC), Lima 15023, Perú.
[2] Dirección de Investigación, Universidad Peruana de Ciencias Aplicadas (UPC), Lima 15023, Perú.

*carlos.raymundo@upc.edu.pe

Abstract. In an environment where the international market demands a large amount of fiber, the countries that produce raw material must be prepared to face the fulfillment of exports. However, it must be noted that in majority of the countries, this fiber, such as alpaca, can be found in the mountains at more than 3,000 MASL where frosts are constant and cause deaths. Likewise, if the collection of feces is not adequate, it can represent an infectious focus which, in addition to the deaths, would reduce the productivity of the fiber. Thus, we can say that the main cause is the incipient knowledge of sanitary practices. This document proposes an alpaca waste management model for the generation of biogas for heating alpacas in frost seasons and obtaining biol for the improvement of pastures. The stages in which the project is carried out include infrastructure of the house, connection of house alpaca and system of drainage, biodigester, improvement of pastures. This model was validated by means of a continuous biodigester prototype supplied with alpaca feces and pig biol achieving 1 of biogas.

1. Introduction

At present, Peru is the main exporter of alpaca fiber, accounting for 84% of all alpaca heads worldwide (INEI, 2012). These exports reached US$ 68.3 million between January and November 2017, representing an increase of 110% over the exports in the previous year, according to the Ministry of Foreign Trade and Tourism (Mincetur). Alpaca production is mainly distributed among the Puno, Cusco, and Arequipa regions, with Puno being the largest producer (65.4% of alpaca heads). The situation is favorable for Puno, as it would become the largest supplier of this type of fiber. However, there is deficit not only in the amount of fiber extracted per alpaca but also in terms of quality. This problem is mainly caused by the incipient knowledge of sanitary practices resulting in inadequate breeding conditions for the survival of the species and for fiber. Notably, these conditions occur in times of frost where the alpaca is threatened by abiotic agents such as the weather and biotic agents like accumulated feces within the sheds. To solve this problem, there are several alternatives for the collection and treatment of manure. Therefore, this research article proposes an alpaca waste management model using a combination of environmental stool treatment techniques supported by industrial engineering tools. This creates a sustainable energy system that solves the problem of

IOP Conf. Series: Materials Science and Engineering **689** (2019) 012010 doi:10.1088/1757-899X/689/1/012010

collecting feces, formation of infectious foci, and generation of biogas to produce a thermally comfortable and clean environment for alpaca breeding using biomass treatment.

2. State of the art

In literature, several models have been identified to address the causes of the emerging awareness of sanitary practices in raising these camelids.

2.1. Manure Management Techniques

The anaerobic digester system has become one of the most dominant techniques in addressing the issue of manure, decreasing the polluting potential of the excrement, and reducing greenhouse emissions [1]. It contributes to the use of organic energy and maintenance and improvement of the fertilizing value of treated products [2]. Ribeiro [3] applied it in a pig farm in Brazil, which is the fourth largest producer and exporter of pork in the world. To ensure compliance with food safety and climate change mitigation, he used a flare biodigester for energy purposes and to improve composting. Exploiting pig waste helps avoid some types of diseases not only while breeding but also in the quality of life of the animals themselves. This system is based on manure management, which increases the economic viability of the anaerobic biodigester by eliminating the construction of traditional manure storage wells subsequent to farm digestion [4]. This system features a unique design based on manure storage pits that directly feed into the biodigester. Then, the post-digested manure (digestate) is stored on the floors of these pig houses. It was found that the farm collected 75% of all manure produced by reducing solids and carbons [4]. The biogas content consisted of 29% CO_2 and 71% CH_4. The gas was used to produce 612,504 MJ of electricity during a stable state period of 24 months [4]. The electricity was valued at $11,450 and fulfilled 50.3% of the demand for pig houses and the manure treatment system. Additionally, the pig house design improved the economic viability of this farm [4].

2.2. Structural Design of Recollection Process

Geoffrey Guest [5] and Chavarria [6] mentions the need to move toward a more sustainable food system, which has led to a general increase in the number of life cycle assessments (LCAs). Therefore, his literature [5] seeks to use a detailed comparative LCA in a dairy production farm located in Ontario, Canada, which recently adopted an active composition system that incorporates a solid/liquid separator with screw press before composting in the container, using automatic composting-manure separator technology to determine environmental advantages and disadvantages [7].

Another solution method used by Vermeulen [7], proposes the implementation of dehydration technology by applying DME to use manure as biogas, since the excess of livestock fertilizer generated on a daily basis and the conventional practices are not sufficient for efficient management. This solution for dealing with excessive livestock manure is a novel idea, which would also reduce pollution. The author also determined the risk of fertilizer self-ignition after the dehydration process.

3. Contribution

3.1. Proposed Model

The alpaca waste management (MGRA) model shown in Figure 1 represents the overall project. To develop each stage of the model, a theoretical version is denoted in Figure 2. This model consists of four fundamental phases and a prior phase. The design was created by adapting environmental engineering tools, such as drainage, anaerobic, and heating systems, which were developed using the PDCA methodology and industrial engineering tools. Likewise, a previous stage called Change Management was included in each stage of the model. The MGRA phases in which an adaptive change was generated are detailed below:

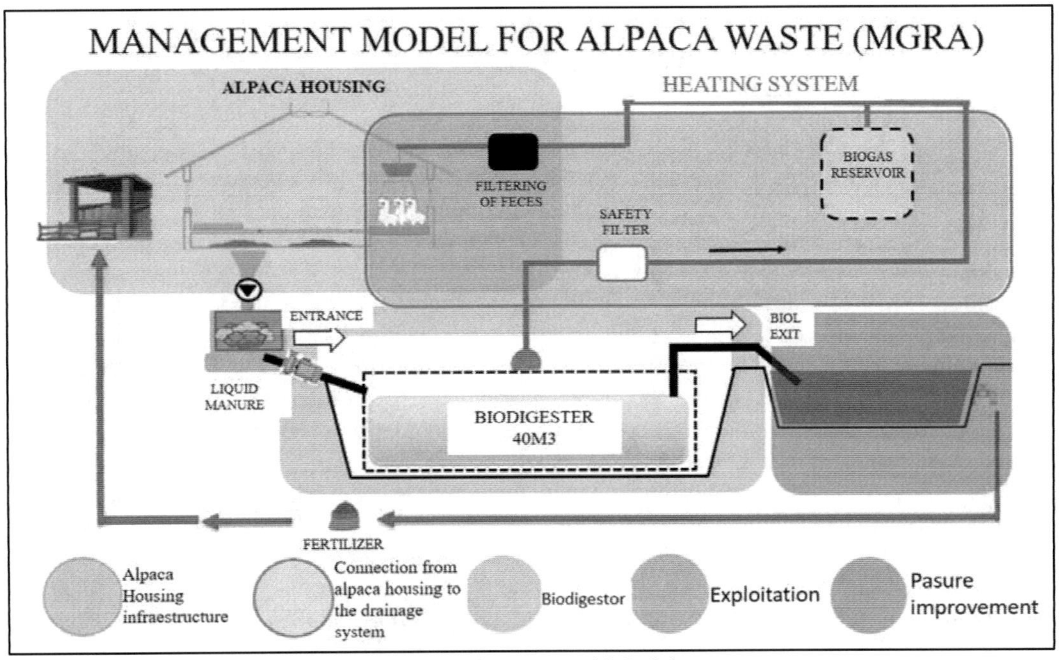

Figure 1. Proposed Model

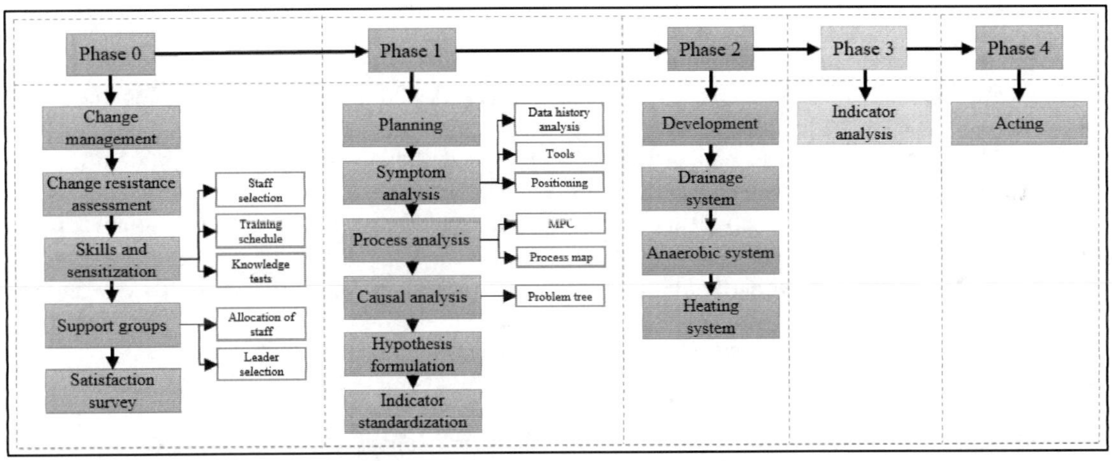

Figure 2. Sequence matrix of the proposal

3.1.1. Phase 0 Change Management

Resistance to Change Assessment: Resistance to change is a key indicator that allows us to assess the level of resistance that the company currently presents. For these purposes, the OCAI method was used because it makes quantifications based on surveys, and the questions are structured by the managers of the organization.

Training and Awareness: The main focus of the training is to convey to farmers the relevance of the problem of low fiber productivity in this area and how they, can develop responsible breeding that ensures optimal production through a comprehensive waste management system. Therefore, it is essential to secure their support for the development and implementation of the proposal.

Support Groups: Support groups are formed at the end of the training. Breeders are selected based on each activity to achieve the set objectives. Likewise, throughout the implementation stage, the trainer will provide 100% support and guide the teams.

Compliance Survey: Measuring the progress of change management will detect the satisfaction level of and rate of resistance to change in employees.

3.1.2. Phase 1 Planning "P1"

Symptom Analysis: The planning stage identified that the productivity of the study site was below average standards. This negative trend is a symptom of the existence of a problem. Subsequently, this trend translates into economic impact reflected in a decrease in income in the region.

Cause Analysis: To identify the root cause, the problem tree is assessed in three large subdivisions, and the main root causes to be resolved are detected.

Hypothesis Formulation: A hypothesis related to the problem found is formulated and the techniques that will resolve it are determined.

Process Standardization: The model processes will be standardized as a working method to serve as an execution guide. To do this, the proposal processes have been mapped, which are analyzed and related to the impact they have on project objectives and customer satisfaction to determine the critical processes. The critical process matrix (MPC) will be applied, along with the process map and flowcharts. This provides an overview of how to manage each process, identify strategic processes and bottlenecks, and understand the importance of each process.

3.1.3. Phase 2 Development "D2–D4"

In this phase, the different systems that allow biomass collection and use are developed using the industrial engineering tools that support the research.

Drainage System: A grid system is proposed along with manure pumping, which avoids surplus digestate storage costs in a housing unit, with a manure storage under the shed. The process begins with the adaptation of the shed (corral) incorporating a storage pit to advantageously collect the stools deposited in the excrement areas located in the center. Once treated, the manure inside the biodigester returns to the pits located in the lower area known as return wells.

Anaerobic System: Biodigesters adapted to the cold are located inside a compact greenhouse with large adobe walls. For the process, it is necessary to take into account the selection of the installation site, considering available spaces and structures, and analyze how to incorporate the system into the sheds. For installation, it will be necessary to consider that the major axis of the digester must be in the direction of east to west, with the greenhouse shed facing the winter sun. In this way, solar radiation can be used.

Heating System: Manure processed under the anaerobic systems serves as fuel for the operation of eight lamps powered by propane or biogas. These are used to provide a thermally comfortable environment for alpacas, which improves the animals' quality of life.

3.1.4. Phase 3 Analysis of "C5"

The indicators are variables that allow for quantitative or qualitative measurement to support the decision-making process. These indices fulfill specific control functions, which are applied according to the needs of each company. During this stage, indicators for each stage of the model are detailed and identified with a code for traceability. Indicators are named based on the following nomenclature: Indicator (First letter of the stage) (Stage No.) - (Indicator No.) MGRA: Name of the indicator

Training and Awareness Indicators: After completing the training and awareness process, a knowledge test will be applied on each trainee in order to determine whether the necessary knowledge was acquired to execute the project. If the person obtains a result equal to or greater than 13, they will be considered approved.

G0-1MGRA Indicator: % Successful Training Sessions

$$\%\,Sucessful\ Trainings\ = \frac{\#\ workers\ with\ grades\ higher\ or\ equal\ to\ 13}{total\ \#\ workers} * 100 \tag{1}$$

Progress Survey Indicators: This indicator shows how satisfied is the person who trained with the project development. Also, this metric helps diagnose discrepancies within the work group.

G0-2MGRA Indicator: % Satisfactory Progress

$$\text{\% Satisfactory Progress} = \frac{Sum\ of\ Grades}{\#\ trainees} * 100 \tag{2}$$

Drainage Systems: In the first phase of the development process, manure is collected seeking to compile the largest amount of biomass possible. To determine if the process is successful, the first indicator must be greater than 90%, and the second must be greater than or equal to 3 kg per day per alpaca.

D2-1MGRA Indicator: % Manure Collected

$$\%\ Manure\ Collected = \frac{Total\ manure\ collected\ (KG)}{Total\ manure(KG)} * 100 \tag{3}$$

Anaerobic Systems: The second phase of the development process is the anaerobic system, where the biomass collected through the drainage system is treated. The objective of this phase is to obtain biogas and biol. To determine if the process is successful, the amount of biogas must be greater than or equal to 0.78 m3 per day. Also, temperature of 19–23°C and pH level of 7.06–7.22 should be maintained.

Indicator D3-1MGRA: Biodigester Efficiency

$$\%\ Biodigester\ Efficiency = \frac{Total\ biogas\ obtained(m3)}{3.6\ m3} * 100 \tag{4}$$

Indicator D3-2MGRA: Average Daily Biodigester Temperature

$$Average\ Daily\ Biogester\ T° = \frac{Sum\ of\ Temperatures\ Recorded}{\#\ Records\ per\ Day} * 100 \tag{5}$$

Heating System: The objective of this phase is to keep the biogas lamps going for approximately 12 hours so that the alpacas achieve a microclimate ensuring their survival in times of frost. The first indicator (D4-1MGRA) provides the average temperature inside the sheds, which should be maintained at 19°C on average. The second indicator (D4-2MGRA) records the consumption of biogas per hour, which must be at least 0.065 m3 per hour. Finally, the D4-3MGRA indicator provides the biogas pressure in contact with the operation of the lamps. This number, according to literature, should be greater than 28 millibars.

D4-1MGRA Indicator: Average Shed Temperature

$$Average\ Shed\ Temperature = \frac{Sum\ of\ Temperatures\ Recorded}{\#\ Records} \tag{6}$$

Indicator D4-2MGRA: Biogas Consumption per Hour

$$Biogas\ Consumption\ per\ Hour = \frac{Biogas\ used\ in\ m3}{\#\ hours\ of\ operation} \tag{7}$$

3.1.5. Phase 4 Acting "A6"

This phase consists of assessing unexpected results. Basically, it is called the feedback stage, as in this phase, process improvement opportunities are identified. Likewise, the PDCA cycle allows improvement assessment to continue, returning to the initial stage in search of new problems that require improvement, thus becoming an endless cycle.

4. Validation

The research study is conducted in Puno, taking into consideration all the variables such as climate, temperature, latitude, pressure, number of alpacas, and other variables used for model simulation. The alpaca owners are small breeders, meaning ordinary people who do not have a company incorporated; therefore, the project is not focused on one company but on the entire sector.

The validation of the MGRA steps are shown below:

Stage 0: Change Management

In the first stage, employees are trained and sensitized to the problems in the sector and the methodology of the management model. At the same time, support groups and progress surveys are created to guarantee staff satisfaction.

Stage 1: Planning "P1"

With the MPC, there was proof that the processes with the greatest impact on objectives included the heating system, production (anaerobic system), and manure collection. Given that the MGRA is a new proposal, which is not included in the current alpaca fiber value chain, the standardization of the processes within this model has been proposed with support from the process map to visualize the overall flow.

Stage 2: Development "D2–D4"

The experimentation of the anaerobic system prototype was performed in three continuous biodigesters. Owing to this simulation, biogas production was evaluated. The simulation of the operation of the anaerobic system was developed in three buckets of 50 liters each, with an initial load of 60 kg of manure.

Once completed, the data collected on Day 30 found production of 0.3 m3/day, thus giving an optimal result for the project. It is also emphasized that the system will produce biogas with a higher concentration of methane for a longer fermentation time. The results of the most important variables of the experiment are shown in table 1.

Table 1. Results Table

Variables	End
Biogas obtained (m3)	0.3
Lamps on	2
Duration of time lamps on	30 min

Stage 3: Assessment of "C5" Indicators

At this stage, the techniques implemented to determine if the proposals have positively impacted the root cause of the problem are analyzed. Table 2 shows the results of the indicators determined for the validation of MGRA.

Table 2. Results of Indicators

Indicator	Coding	Reference level	Results	Validation
Percentage of Successful Trainings	G0-1MGRA	90% - 100%	90%	complies
Percentage of Satisfactory Progress	G0-2MGRA	90% - 100%	90%	complies
Percentage of Manure Collected	D2-1MGRA	90% - 100%	*	
Percentage of Manure Collected Daily	D2-2MGRA	90 - 150 kg	*	
Average Daily Biodigester Efficiency	D3-1MGRA	90% - 100%	70%	does not comply
Biogester Temperature	D3-2MGRA	19 - 23 °C	20 °C	complies
Biodigester pH Level	D3-3MGRA	7.06 - 7.22	7.2	complies
Average Shed Temperature	D4-1MGRA	19 - 30 °C	19 °C	complies
Biogas Consumption Per Hour	D4-2MGRA	0.0325 - 0.065 m3	0.060 m3	complies
Biogas Pressure	D4-3MGRA	20 - 28 millibars	23 millibars	complies
Ppm of Hydrogen Sulphide	D4-4MGRA	0 ppm	**	

Stage 4: Acting "A6"

At this stage, the results obtained from the MGRA proposal are assessed. However, another run of the model could not be executed because of it being a real experimentation on a prototype scale.

IOP Conf. Series: Materials Science and Engineering **689** (2019) 012010 doi:10.1088/1757-899X/689/1/012010

5. Results

- 90% of the breeders polled agreed with the implementation of the model. This aspect is favorable for better model implementation.
- Training progress of 90% was achieved.
- An efficiency of 70% in the operation of the biodigesters was achieved.
- The biodigesters were maintained at a temperature of 20°C to facilitate decomposition.
- A pH of 7.2 was achieved, which is optimal according to the literature.
- A temperature of 19°C was achieved inside the shed. (The shed was simulated, and the temperature was taken with the lamps switched on for 30 minutes).

6. Conclusions

This research project managed to compare and combine environmental tools using the PDCA methodology and industrial engineering tools within the alpaca waste management (MGRA) model. Additionally, the experimentation of the three prototypes of biodigesters was important in demonstrating the extraction of biogas from alpaca manure. It should be noted that estimated values calculated from literature were considered through drainage system concepts, amount of manure per day, number of total solids, retention time, biodigester volume, and biogas production, among other variables.

The prototype development used an initial mixture of biomass to fill a bucket with 12 kg of alpaca manure and 36 liters of pork boil at a ratio of 1 to 3. In this way, 0.3 m3 of biogas of could be obtained per day in a production span of 30 days, which implies that the estimated amount of biogas will be reached (0.78 m3 per day to keep eight lamps switched on for 21 days).

The reduction in alpaca mortality could not be directly validated. However, this could be inferred, as it was possible to integrate a heating system to guarantee a microclimate of 19°C inside the shed. In the same way, manure collection could not be validated either, since it would be necessary to implement the full-scale model. However, the dimensions of the shed were conditioned according to the literature, and a design for the alpacas shed was proposed.

Finally, the project standardized the MGRA processes so that, whenever required, each step can be carried out within the process map and the critical process flowcharts.

References

[1] Braber K. Anaerobic digestion of municipal solid waste: a modern waste disposal option on the verge of breakthrough. Biomass Bioenerg 1995;9:365e76.

[2] Arévalo, J., Quispe, G., Raymundo, C., "Sustainable Energy Model for the production of biomass briquettes based on rice husk in low-income agricultural areas in Peru.", Energy Procedia, 141, pp. 138-145.

[3] Ribeiro, E.M.a, Mambeli Barros, R.bEmail Author, Tiago Filho, G.L.b, dos Santos, I.F.S.b, Sampaio, L.C.c, dos Santos, T.V.c, da Silva, F.D.b, Silva, A.P.M.b, de Freitas, J.V.R.d (2018). Feasibility of biogas and energy generation from poultry manure in Brazil . Waste Management and Research. Pages 221-235.

[4] Robert J. Meinena, Kenneth B. Kepharta, Robert E. Gravesb (2014). Economic feasibility and evaluation of a novel manure collection and anaerobic digestion system at a commercial swine finisher Enterprise. Biomass and Bioenergy Pages 10-21

[5] Geoffrey Guest, Ward Smith, Brian Grant, Andrew VanderZaag, Ray Desjardins y Brian McConkey (2017). A comparative life cycle assessment highlighting the trade-offs of a liquid manure separatos-composter in a Canadian dairy farm system. Journal of Cleaner Production . Pages 824-835.

[6] Chavarria, D., Ramos, R., Raymundo, C.," Development of a hybrid heating system based on geothermal–photovoltaic energy to reduce the impact of frosts on inhabitants of rural areas in the ring of fire, southern Peru", Smart Innovation, Systems and Technologies 140, pp. 131-139.

[7] Vermeulen J, Huysmans A, Crespo M, Van Lierde A, De Rycke A, Verstraete A. Processing of biowaste by anaerobic composting to plant growth substrates. In: Proceedings of the

international symposium on anaerobic digestion of solid waste, Venice; 14e17 April, 1992. pp. 147e57.

IOP Conf. Series: Materials Science and Engineering **689** (2019) 012011 doi:10.1088/1757-899X/689/1/012011

Fast Switching Valve Utilization to Control Pneumatic Cylinder Speed

M. Adel Elkashef[1], Mohamed Aly[2], Osama E. Mahmoud[2], and Yehia H. Hossamel-din[3]

[1] Department of Mechatronics Engineering, HTI, Egypt.
[2] Mechanical Power Engineering, Faculty of Engineering-mataria, Helwan University.
[3] Chairman of Mechanical Engineering Department, Faculty of Engineering and Technology, Future University in Egypt (FUE).

Email: Hamoda_adel@hti.edu.eg, Mechhanafy@gmail.com, Osama_ismail@m-eng.helwan.edu.eg, Yehia.hendawy@fue.edu.eg

Abstract. The position and speed of the piston of pneumatic cylinder are very important parameters in evaluating the pneumatic system performance. Utilization of Pulse Width Modulation (PWM) technique to control the Fast Switching Valve (FSV) usually causes fluctuation in the piston motion and accordingly, the system performance deteriorates. The objective of this work is to introduce a novel PWM control technique considering the suitable duty cycles according to the switching frequency in order to improve the FSV characteristics and reduce piston speed fluctuation. A test rig for a simple pneumatic circuit with an FSV as a control element has been designed and realized. The displacement of the piston is measured using a Linear Wire Potentiometer Transducer (LWPT), while its derivative is used to estimate the speed considering a low pass filter to reduce the effect of piston position signal noise. A nonlinear mathematical model is introduced and experimentally validated. The results of both computer simulation and experimental measurements show that the integration between fast switching valve and PWM control signal leads to a quasi-linear relationship between duty cycle and piston speed. Moreover; results proved the successful control of piston speed based on the correct duty cycle and valve switching frequency.

Keywords: Fast Switching Valve, Pulse Width Modulation, Pneumatic Cylinder, PWM Scheme.

1. Introduction

Pneumatic systems are widely used in robotics and industry applications. They are clean, cheap, simple in design and control, self-cooling, fast acting and low maintenance cost. This makes pneumatic systems a preferable choice for many cases compared to electro-mechanical and hydraulic systems [1]. The non-linear behavior of pneumatic systems mainly caused by air compressibility, is a major drawback of pneumatic systems. Therefore, proper control of pneumatic actuators becomes difficult to be achieved. Continuous acting Servo Valves (SV) are used to realize servo-pneumatic systems, but those valves are complex in structure, expensive (about $1000 US), and need special controllers. Many researchers developed the servo-pneumatic systems by using digital valves with fast switching electronic circuit (about $ 90 US) and driven by pulse width modulation technique instead of servo valves. This is in order to decrease system's complexity and cost [2, 3].

PWM-Driven electro-pneumatic systems motivate many researchers to investigate and develop its characteristics in order to improve its dynamic performance. Many studies investigated and developed the modeling and control of servo-pneumatic system including force element (actuator) and command device (valve) [4, 5].

The improvement of dynamic performance of pneumatic system mainly associated with the enhancement of the pneumatic valve characteristics. M. Pipan et al. [6] studied a PWM control algorithm for a closed-loop volumetric flow of a fast switching valve on the basis of experimental results. Topcu et al. [7] investigated the characteristics of 2/2-way fast switching valves and presented a method to reduce closing time. Taghizadeh et al. [8] developed a non-linear dynamic model of 3/2-way fast switching valve (FSV) and validated it experimentally. Shiee et al. [9] used different schemes to investigate the performance of the servo-pneumatic system. Taghizadeh et al. [10] investigated the dynamics performance of PWM-Driven servo-pneumatic system and presents an open loop behavior between the input voltage and the speed of piston rod.

In this paper, a new simple method is investigated to control the speed of cylinder by using the fast switching valve. The effect of control signal pulse width on the speed of cylinder is investigated theoretically and experimentally.

2. Experimental Setup

Referring to Fig. 1, a double acting cylinder actuator (DNSU-20-100-PPV-A) from FESTO including stroke length of 100 mm, and 20 mm bore diameter is used to be under control. FESTO fast switching valve 3/2-way is used as a command device to control the air flow. A manual flow control valve is used to blank the exhaust port or connect it to the atmosphere. An NI-myRIO controller is used to control the Fast Switching Valve (FSV). The control signal of the FSV is transferred through PWM output pin of the I/O port toward the MOSFET amplification circuit. HONTKO Linear Wire Potentiometer Transducer (LWPT) is attached to the end of piston rod to measure its displacement. This signal is transferred through the analog input port of the NI-myRIO to be recorded and monitored. A reciprocating air compressor with enough air tank capacity is used to supply pressurized air of 6 bars. A service unit from FESTO (FRC-M7-D-7-5M) is used. This unit contains filtration, regulation, and lubrication stages in a single unit with pressure gauge. The air pressure is regulated through the service unit to be 5 bars for all experiments.

3. The Proposed Pneumatic Circuit

The proposed pneumatic circuit is shown in Fig. 2. As can be seen from the figure, the full diameter side chamber is connected to the output port of the valve, and the annulus side chamber is connected directly to the service unit. The input port of the valve is connected to the regulated pressure while the exhaust port is connected to the flow control valve. Energizing of the solenoid allows the pressurized air to pass through the valve toward the full side chamber which builds up the pressure inside it. The self-return spring of the valve returns the spool and maintains its normal position when the solenoid is de-energized. The higher switching operating frequency of the valve result in smoother displacement of the piston rod and bigger pulse duty cycles result in bigger packets of air. The air packets delivered to cylinder have a high rate of delivery which is faster than the dynamics of the cylinder, so that, the system filters this discreteness of air packets and piston rod responds to the average of this packets.

The extension of the piston requires to blank or to throttle the exhaust port of the fast switching valve by adjusting the flow control valve. The flow control valve is completely closed during the experiments carried out in the present work and it is only opened to retract the piston when the fast switching valve is not energized. The extension speed of the piston rod is controlled by the FSV control signal.

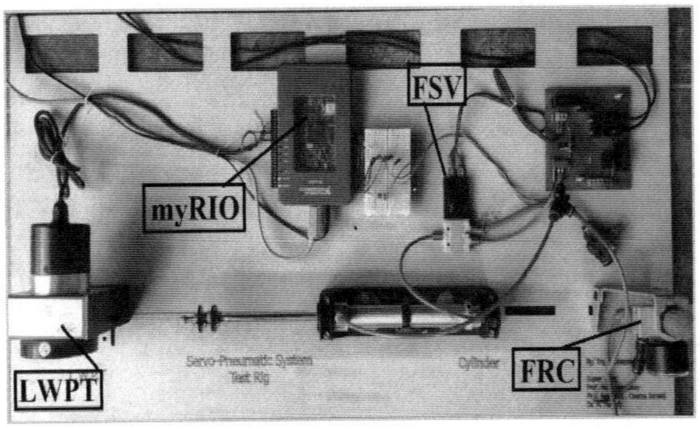

Figure 1. Photograph of Test Rig

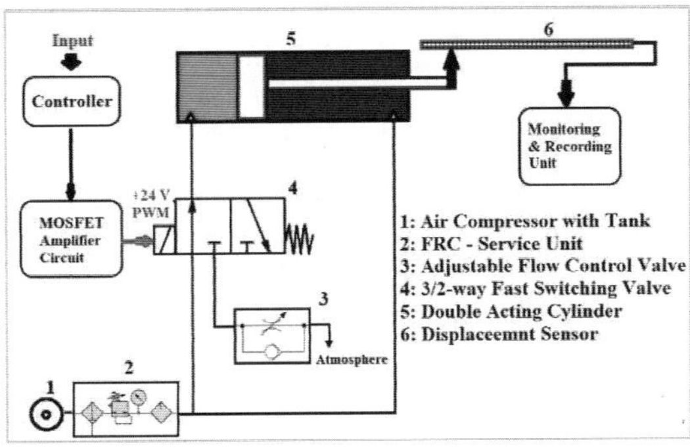

Figure 2. The Proposed electro-pneumatic circuit

4. Modeling and Simulation

In this section, the mathematical equations governing the proposed system are introduced. Moreover, the validation is carried out using the experimental setup described in section 2.

4.1 Mathematical equations

A non-linear dynamic model of the proposed pneumatic circuit is presented including the force element and the command devices. The system can be represented by the following set of differential equations:

- The equation of motion for the pneumatic cylinder:

Applying Newton's second law, the following equation can be obtained:

$$\frac{d^2x}{dt^2} = \frac{1}{m_L}\left[P_1 A_1 - P_s A_2 - b\frac{dx}{dt}\right] \qquad (1)$$

where, x is the displacement of piston rod, m_L is the mass of piston and load, P_1 is the pressure inside the full side chamber, P_s is the supply pressure connected to annulus side chamber. A_1 is the full side piston area, A_2 is the annulus side piston area, and b is the cylinder viscous coefficient.

- The equation of rate of change for pressure inside full side chamber:

$$\dot{P}_1 = \frac{RT}{V}\left(\propto_{in} \dot{m}_{in} - \propto_{out} \dot{m}_{out}\right) - \propto \frac{P_1}{V}\dot{V} \qquad (2)$$

where, $\propto_{in}$, *and* $\propto_{out}$ are taking values between 1 and 1.4 depending on the actual heat transfer during the process. V is the control volume, T is the temperature of air and R is the ideal gas constant. $\dot{m}_{in}$ and $\dot{m}_{out}$ are the mass flow rate entering and leaving the chamber.

- The mass flow rate through valve orifice of area A_v can be expressed by:

$$\dot{m} = \begin{cases} C_f C_1 A_v \frac{P_u}{\sqrt{T}}, & \frac{P_d}{P_u} \le 0.528 \\ C_f A_v C_2 \frac{P_u}{\sqrt{T}} \left(\frac{P_d}{P_u}\right)^{1/k} \sqrt{1 - \left[\frac{P_d}{P_u}\right]^{\frac{(k-1)}{k}}}, & \frac{P_d}{P_u} > 0.528 \end{cases} \qquad (3)$$

where, C_f is the discharge coefficient, P_u is the supply pressure, P_d is the downstream pressure, T is the temperature of supplied air (293 °K), and k is the specific heat ratio. C_1 = 0.04042, and C_2 = 0.15617 are constants of a given fluid [11]. MATLAB/SIMULINK environment is used to simulate the above dynamical equations of the system. It should be noted that the effect of cushioning at the annulus side of the piston has been considered by variable spring and damper coefficients, while it is neglected at the full side.

4.2 Model validation

The validation of the mathematical model for the proposed system is carried out by comparing the results of the system simulation to the corresponding measured ones. Fig. 3 shows the piston displacement obtained from both simulation output and experimental measurements considering 50% duty cycle & 60 Hz PWM valve control signal. Inspecting this figure, it can be seen that there is a very good quantitative and qualitative agreement of both simulation and experimental results (maximum peak to peak error = 4.85% of full displacement stroke).

In this work, the piston speed is estimated from its displacement, it could not be obtained from the direct differentiation of the displacement signal as it was too noisy. Therefore, Digital low pass filter is used to alleviate this problem. Fig.4 shows the piston speed obtained from both simulation output and experimental measurements at the same duty cycle and frequency (50% duty cycle & 60 Hz PWM valve control signal).

It is clear from the figure that there is a good qualitative agreement between both simulation and experimental results. However; there is still a noticeable but acceptable oscillation in the measurement results which, as mentioned before, is due to the fact that the speed is estimated from the measured displacement. Accordingly, to get more accurate results, it is recommended to measure the speed directly using a speed sensor such as a linear encoder.

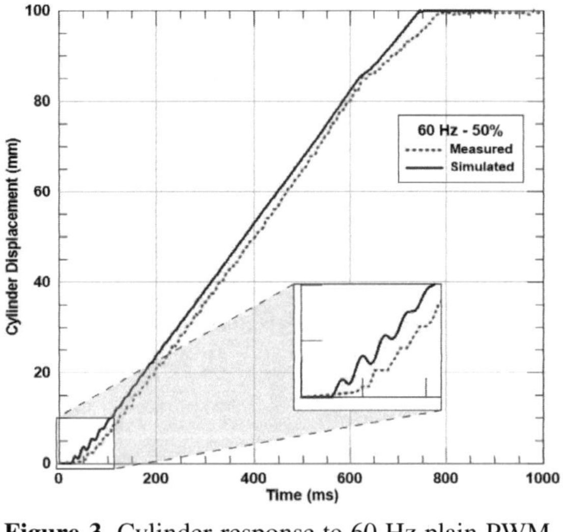

Figure 3. Cylinder response to 60 Hz plain PWM valve control signal.

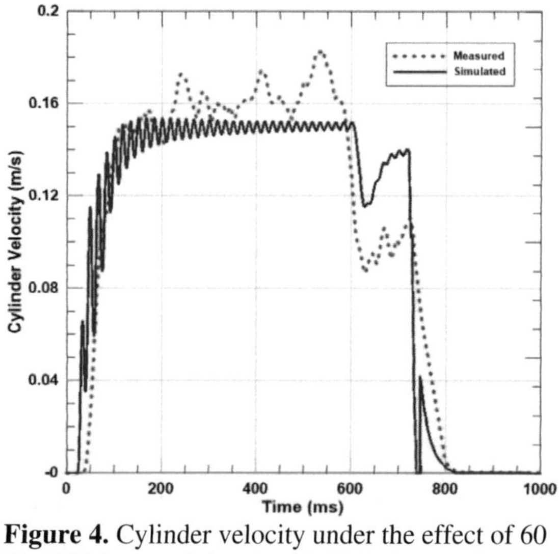

Figure 4. Cylinder velocity under the effect of 60 Hz PWM control signal with 50% duty cycle

5. PWM Scheme

In general, the fast switching valve has a delay opening time (TON) and a delay closing time (TOF). The values of TON & TOF depend on the manufacturing tolerance and the electromagnetic characteristics of the valve solenoid. The existence of those delay times causes a dead band and a saturation band. Accordingly; the valve cannot completely open with duty cycles less than a minimum (DC_{Min}) which corresponds to pulse ON time enough to compensate for the ON delay time. Also, the valve is saturated with duty cycles bigger than a maximum (DC_{Max}) which corresponds to pulse OFF time enough to pass the OFF-delay time. The values of the DC_{Min}, and the DC_{Max} depend on the control signal frequency and the delay times of the valve. Accordingly, the minimum and maximum duty cycle can be given as follows:

$$DC_{Min} = \frac{TON}{T_{PWM}} \quad , \quad DC_{Max} = \frac{T_{PWM} - TOF}{T_{PWM}} \tag{4}$$

Where, TPWM, TON, & TOF are the valve switching period time, delay opening time & delay closing time; respectively. According to the manufacturer, the maximum switching frequency for this FSV is 330 Hz but the ambient temperature must be limited with switching frequencies in excess of 125 Hz. Also, this valve has a delay opening time, TON=1.7 ms, and a delay closing time, TOF=2 ms. Considering equation 4, DC_{Min} & DC_{Max} are calculated and Table 1 shows the maximum and the minimum duty cycles for the FSV operating frequencies used through this study.

6. Open Loop System Behavior

The open loop behavior of the proposed system is investigated by simulating the model with several duty cycles at different frequencies considering constant pressure supply of 5 bars. Using the developed validated simulation model, the following are to be discussed:

- The pressure in the piston full diameter side chamber and the valve spool position.
- The piston speed time response.

6.1 The pressure in the piston full diameter side and the valve spool position

Fig. 5 shows the relationship between the pressure inside the full side chamber, and the displacement of valves' spool under the effect of 50% duty cycle & 60 Hz PWM control signal.

The PWM valve control signal is applied to start switching of the valve at t = 0 which causes pressure to build up inside the full piston diameter side chamber. After the pressure reaches certain value which overcomes the opposing forces, the piston starts to move causing pressure drop. The new air packet compensates for the pressure drop and it builds up again causing the piston to continue motion.

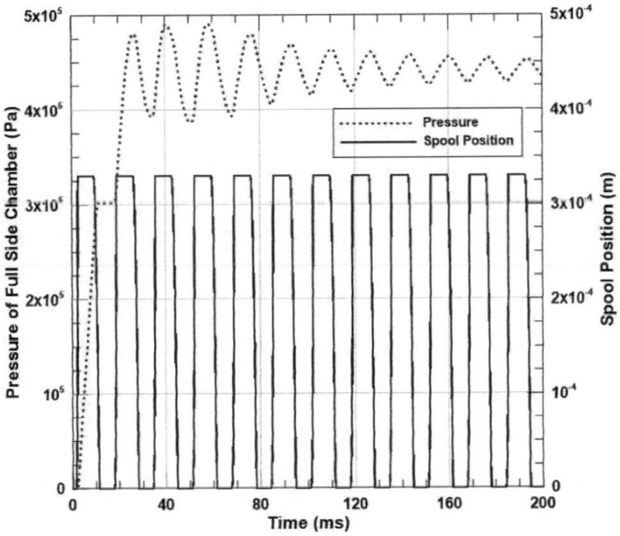

Figure 5. The relation between the pressure inside the full side chamber and the position of spool.

Table 1. limits of duty cycles

Frequency	Min Duty Cycle	Max Duty Cycle
60 Hz	0.1	0.88
80 Hz	0.136	0.84
100 Hz	0.17	0.8

6.2 The piston speed time response

Examples for piston speed time response are shown in Fig. 6 for valve switching frequencies of 60 Hz. The results shown in Fig. 6 indicate the following:

- The stroke of cylinder can be described by three sequential stages, namely: dead zone, fluctuation zone, and smooth motion zone. Both of experimental and simulated results indicated that the larger duty cycles result in smaller dead zone while the higher PWM frequencies and cushioning result in smaller fluctuation zone and larger smooth zone.

- The speed of cylinder can be controlled by the pulse width of the valve control signal.
- The switching frequency of FSV limits the maximum and minimum duty cycles in order to avoid both valve saturation condition (the valve remains completely open) and the unpredicted long dead band.
- In the conditions considered in Fig. 6; as duty cycle increases, the cylinder speed increases till DC_{Max}. For duty cycles above the DC_{Max}, the speed can be considered the same (90 % & 95% duty cycles).
- The use of higher switching operating frequency for the FSV results in a smoother motion of the piston rod.

The cylinder speed versus the control signal duty cycle for three different valve operating frequencies (60 Hz, 80 Hz, and 100 Hz) is plotted in Fig. 7 from which the following can be noticed:

- The input-output relationship is a quasi-linear one.
- The use of higher operating frequency decreases the range between the maximum and the minimum duty cycle (increases DC_{MIN} & decreases DC_{MAX}) which is in agreement with equation 4.
- The piston speed is reduced with the use of the 10% duty cycle of both 80 Hz & and 100 PWM control signal compared to the same duty cycle of 60 Hz PWM control signal as it lies below the minimum limits of duty cycle for both frequencies.

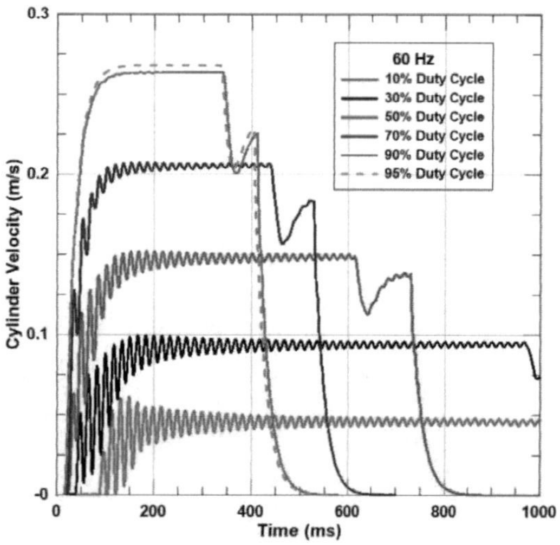

Figure 6. The velocity of simulated cylinder under the effect of 60 Hz PWM control signal.

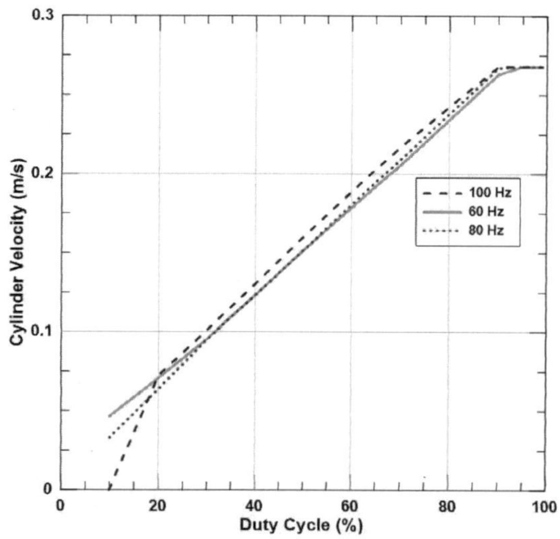

Figure 7. The quasi-linear input-output behavior

7. Concluding Remarks

In this work, a reliable test rig for simple pneumatic circuit including an FSV as a command element has been designed and built. Also, a mathematical model has been introduced and validated using this test rig. This is to investigate the usage of PWM in controlling a fast switching valve as a part of an electro- pneumatic system. From both experimental and simulation results obtained; it can be concluded that a) utilization of PWM technique to control the fast switching valve reduces the non-linearity behavior of the valve, b) the fast switching valve can be used effectively to control the speed of pneumatic actuators, c) direct differentiation of displacement signal to obtain speed signal may not give correct results due to the displacement signal noise and low pass filter may be implemented to alleviate this problem, d) the bigger duty cycles result in higher speed of piston and smaller dead zone, and e) a simple algorithm can be used to avoid both dead band and saturation.

8. Future Works

Investigate, theoretically and experimentally, the optimal operating frequency of the 3/2-way fast switching valve for controlling a servo-pneumatic system.

References

[1] P. Scarfe, and E. Lindsay: Air muscle actuated low cost humanoid hand, International Journal of Advanced Robotic Systems, 3 (2006) p. **22**.

[2] M.B. Thomas: Advanced servo control of a pneumatic actuator, PhD Thesis, The Ohio State University, 2003.

[3] G.M. Bone, M. Xue, and J.J.M. Flett: Position control of hybrid pneumatic–electric actuators using discrete-valued model-predictive control, Mechatronics Journal, 25 (2015) p. **1-10**.

[4] B. Najjari, S.M. Barakati, A. Mohammadi, and M. Bostanian: Position control of an electro-pneumatic system based on PWM technique and FLC, ISA transactions, 53(2014) p. **647-657**.

[5] H. Fawaz, and M.A. Aziz: Position control of a pneumatic actuator using digital valves and fuzzy PI controller, International Journal of Mechanical and Mechatronics Engineering, 16(2016) p. **10-22**.

[6] M. Pipan, and N. Herakovic: Closed-loop volume flow control algorithm for fast switching pneumatic valves with PWM signal, Control Engineering Practice Journal, 70 (2018) p. **114-120**.

[7] E.E. Topçu, İ. Yüksel, and Z. Kamış: Development of electro-pneumatic fast switching valve and investigation of its characteristics, Mechatronics Journal, 16 (2006) p. **365-378**.

[8] M. Taghizadeh, A. Ghaffari, and F. Najafi: Modeling and identification of a solenoid valve for PWM control applications, Comptes Rendus Mecanique Journal, 337 (2009) p. **131-140**.

[9] M. Shiee, A. Sharifi, M. Fathi, and F. Najafi: An experimental comparison of PWM schemes to improve positioning of servo pneumatic systems, The International Journal of Advanced Manufacturing Technology, 82 (2016) p. **1765-1779**.

[10] M. Taghizadeh, A. Ghaffari, and F. Najafi: Improving dynamic performances of PWM-driven servo-pneumatic systems via a novel pneumatic circuit, ISA transactions, 48 (2009) p. **512-518**.

[11] E. Richer, and Y. Hurmuzlu: A high performance pneumatic force actuator system: Part I-Nonlinear mathematical model, Transactions-American Society of Mechanical Engineers Journal of Dynamic Systems Measurement and Control, 122 (2000) p. **416-425**.

Numerical and experimental analysis of the rotor eccentric effect on the labyrinth seal

Jintao Mo[1], Zhiwei Yu[1], Ying Luo[1], Hua Du[1]
[1]Nuclear Power Institute of China, Chengdu, PR China

E-mail: mojintao@126.com Phone: 028-85908790

Abstract. Labyrinth seals are widely used in turbomachines. In this paper, a transit simulation method is introduced to model the relationship between the journal offset in Labyrinth seals and the system stability based on Fluent UDF. A special dynamic mesh method has been adopted to ensure fine mesh quality during the arbitrary motion of the journal. In each time step, the transient fluid seal force on the journal surface is obtained by the solving Navier-Stokes equations. Then the displacement of the journal in the next step can be numerically solved by the FEM model of the rotor system. For every transient time step, the computed fluid seal force and the rotor motion displacement were coupled by data exchange. The experiment is carried out in a steam turbine with a position adjustable labyrinth seal. The experimental results show that the amplitude of the vibration decreases after the change of the relative position between the journal and the labyrinth seals and the numerical results show that the locus of the journal under different labyrinth seal offsets has some half-frequency components. Both the simulation and the experiment validates that the relative position between the journal and the labyrinth seals can greatly affect the stability in the coupled rotor-seal system.

1. Introduction

Labyrinth seals, which are widely used in turbomachines to reduce leakage, has close relationship with the stability of the rotor system due to self-excited vibrations induced by the seal fluid force under certain conditions. Therefore, it is necessary to evaluate the forces in labyrinth seals in a precise rotor dynamic analysis. A linear model is first introduced in reference [1-3] to illustrate the relationship between the motion of the journal and the seal fluid force. The general form can be seen in Eq(1). If the journal moves in a small area, the seal fluid force is proportional to the displacement as well as the velocity of the rotor and adequately accurate solution can be got under the condition. However, if the vibration amplitude is large, this linearized equation of motion will fail to give reliable results.

$$-\begin{Bmatrix} F_x \\ F_y \end{Bmatrix} = \begin{bmatrix} K & k \\ -k & K \end{bmatrix} \begin{Bmatrix} X \\ Y \end{Bmatrix} + \begin{bmatrix} C & c \\ -c & C \end{bmatrix} \begin{Bmatrix} \dot{X} \\ \dot{Y} \end{Bmatrix} \tag{1}$$

$$-\begin{Bmatrix} F_x \\ F_y \end{Bmatrix} = \begin{bmatrix} K - m_f\tau^2\omega^2 & \tau\omega D \\ -\tau\omega D & K - m_f\tau^2\omega^2 \end{bmatrix} \begin{Bmatrix} X \\ Y \end{Bmatrix} + \begin{bmatrix} D & 2\tau\omega m_f \\ -2\tau\omega m_f & D \end{bmatrix} \begin{Bmatrix} \dot{X} \\ \dot{Y} \end{Bmatrix} + \begin{bmatrix} m_f & 0 \\ 0 & m_f \end{bmatrix} \begin{Bmatrix} \ddot{X} \\ \ddot{Y} \end{Bmatrix} \tag{2}$$

A nonlinear dynamic model [4-6] is then introduced as the Eq(2) shows. The equation couples the fluid film radial stiffness, damping and inertia effects, and has been applied in many calculations [7-9]. However, the accurate solution of these equation requires reliable coefficients. Although the rotordynamic coefficients of labyrinth seals have been calculated or tested by many researchers [10-12], empirical corrections are still needed to allow calculations matching the experimental results because the detailed structure can not be included in the model. This greatly limits the application of these models.

With the development of the computer technology, more and more researchers chose to use CFD to calculate the seal fluid force and leakage in Labyrinth seals [13-16]. As more geometric details can be considered in CFD, more reliable results can be got.

In this paper, in order to simulate the transit flow field in a labyrinth seal, the commercial CFD software Fluent and a dynamic mesh method similar in journal bearing calculation [17] are adopted. The method can ensure fine mesh quality during the arbitrary motion of the journal so that the transit analysis of the coupled rotor-seal system with an offset labyrinth seal becomes feasible.

2. Method

In order to simulate the effect of the journal offset in the labyrinth seal, the coupled rotor-seal system is built numerically. The system involves two parts. The CFD model is used to calculate the seal fluid force on the journal. The single disk FEM model is used to calculate the transit displacement of the rotor.

2.1. CFD model

The Figure 1 shows the geometric model and the mesh model of the labyrinth seal, they are both built in the GAMBIT. From the mesh section of the geometric, it is obvious that the mesh can be divided into two parts, the chamber part and the clearance part. In the simulation, the clearance part is used as the dynamic mesh zone. This means the UDF macro DEFINE_GRID_MOTION is applied in this area to ensure fine mesh quality during the arbitrary motion of the journal. The detailed dynamic method description can be seen in reference[17] and is not given here for brevity. The number of elements corresponds to the Figure 1 is listed in the Table1.

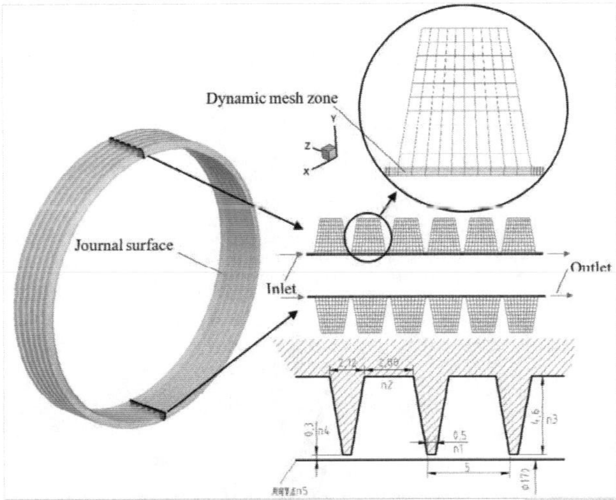

Figure 1. The Geometric dimensions and the mesh of the labyrinth seal

Table 1. The number of elements

Label	Number of elements
n1	5
n2	10
n3	10
n4	5
n5	500

The *k*-*ε* model is selected to simulate the turbulent flow in the chamber. The inlet total pressure is specified to be 1MPa and the outlet static pressure is specified to be 101325Pa. The back flow turbulent intensity and viscosity ratio are specified to be 4% and 4 separately in both the inlet and

outlet. For a transient simulation, as the center of the journal changes over time, the UDF macro DEFINE_PROFILE is applied to specify the rotational speed of the journal according to the reference [17].
The SIMPLEC method is selected to solve the discrete equations. The first order implicit format is chosen as the temporal discretization format. The detailed spatial discretization formats are listed in the Table2.
The time step is 0.0001s and the iterations in each step is 20.

Table 2. The spatial discretization format

Item	Spatial discretization format
Gradient	Least squares cell based
Pressure	Standard
Density	Second order upwind
Momentum	Second order upwind
Turbulent kinetic energy	First order upwind
Turbulent dissipation rate	First order upwind
Energy	Second order upwind

2.2. Coupled rotor-seal system
The Figure 2 shows the single disk model, the length of the axis is 0.63m and the mass of the axis is ignored. The radius of the axis is 0.03m. The weight of the disk is 0.7351kg and the eccentric distance on the disk is 1.2mm. The bearing clearance is 3e-4m and the bearing length is 0.3m.

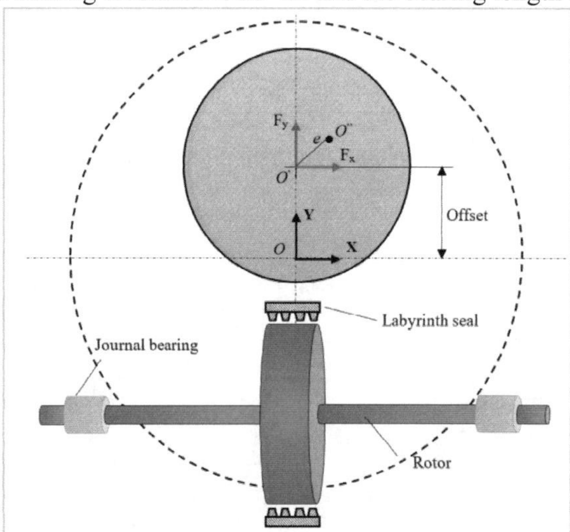

Figure 2. The rotor model description

The equation of the rotor system can be seen in Eq(3).

$$\begin{bmatrix} M & 0 \\ 0 & M \end{bmatrix}\begin{Bmatrix} \ddot{X} \\ \ddot{Y} \end{Bmatrix}+\begin{bmatrix} C & 0 \\ 0 & C \end{bmatrix}\begin{Bmatrix} \dot{X} \\ \dot{Y} \end{Bmatrix}+\begin{bmatrix} K & 0 \\ 0 & K \end{bmatrix}\begin{Bmatrix} X \\ Y \end{Bmatrix}=\begin{Bmatrix} F_x \\ F_y \end{Bmatrix}+\begin{Bmatrix} f_x \\ f_y \end{Bmatrix}+\begin{Bmatrix} Q_x \\ Q_y \end{Bmatrix} \quad (3)$$

Where M is a mass matrix; C and K are the damping matrix and stiffness matrix respectively; F is the oil film force calculated by the short bearing theory; f is the seal fluid force calculated by the CFD model in Fluent; Q is the unbalanced force.
The whole coupled rotor-seal system calculation procedure can be seen in the Figure 3. In order to consider the journal offset in the seal, before the transient calculation, a static calculation with the given offset should be done as the initialization for the following simulation. The detailed description of the relative position between the journal and the seal can be seen in the Figure 2. Then, like the flow diagram shows, in every time step, the UDF macro first calculate the seal fluid force on the journal

surface to write into the .txt files and call the Matlab code to process the generated files. The Matlab programe reads in the seal fluid force and begins to solve the Eq(3). After the Newmark integration, the Matlab code writes the displacement for the temporary time step into .txt files. Then, when the UDF code judge the Matlab program has been successfully executed, it reads in the disk displacement to update the mesh in DEFINE_GRID_MOTION macro and also update the journal surface velocity in DEFINE_PROFILE macro. After all of these have been done, the Fluent records all the data you want and starts to iterate. When the iterations of the temporary time step is over, if the iteration demand is not fulfilled, the above procedure will be repeated, else the whole calculation will be ended.

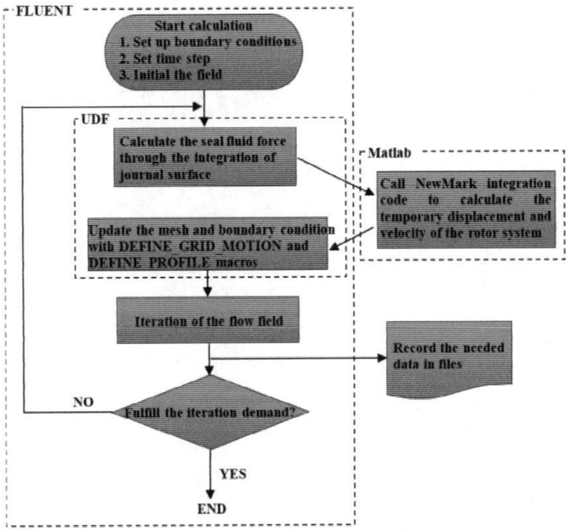

Figure 3. The flow diagram of the coupled rotor-seal system

3. Calculation results

The whole calculation results can be divided into two parts. The first part focuses on the relationship between the rotor-seal system stability and the journal offset. The second part reveals the transit characteristics of the seal fluid field under different axis orbits.

3.1. Stability description

The Figure 4 shows the rotor axis orbits under different journal offset, obviously, with the change of the offset, the axis orbits under different offsets are quite different. When the offset is 0μm and 100μm, the axis orbit is a circle, its amplitude is influenced by the unbalanced force. Ordinarily, this kind of axis orbit is considered stable. However, when the offset is -80μm and -100μm, the half-frequency vibration can be seen from the axis orbits. This kind of axis orbit is viewed unstable. What's more, when the offset shifts from -80μm to -100μm, the half-frequency vibration seems more obvious. So it is easy to draw the conclusion that the relative position between the journal and the labyrinth seals can greatly affect the stability in the coupled rotor-seal system. Often, especially the huge turbinemachine rotor systems can easily become unstable due to the bad uniformity of seal fluid force in the labyrinth seals. So if the seals relative position with the rotor can be adjusted properly, the whole coupled rotor-seal system stability may be improved.

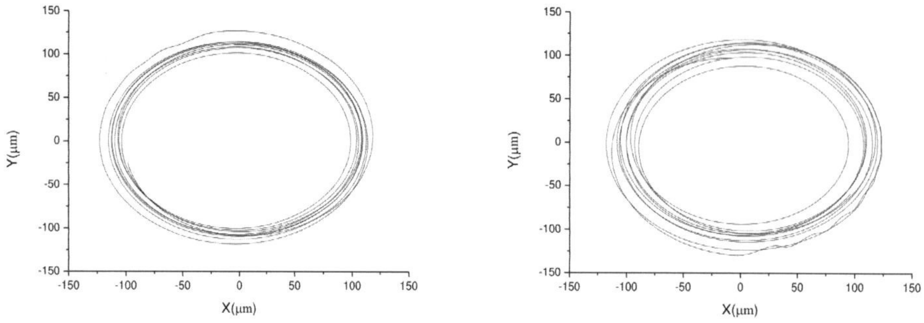

(a)Offset=100μm **(b)Offset=0μm**

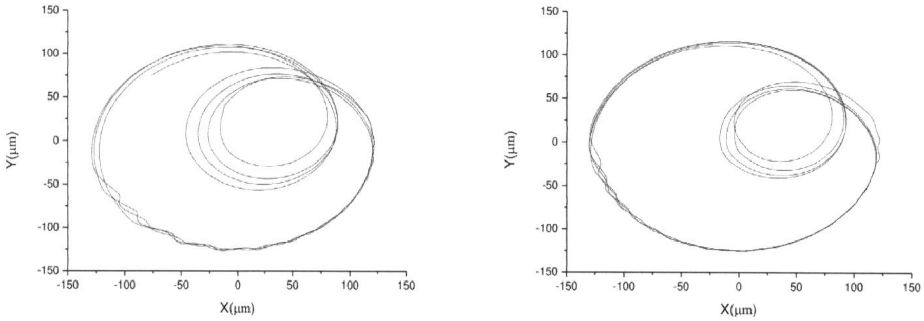

(c)Offset=-80μm **(d)Offset=-100μm**

Figure 4. The axis orbits under different journal offset

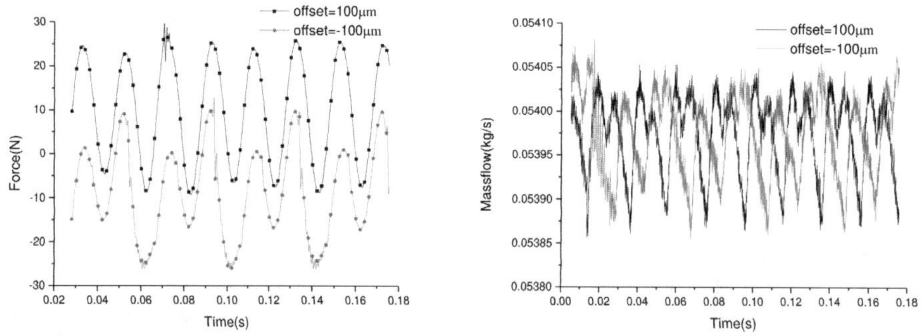

(a) seal fluid force **(b) leakage**

Figure 5. The transit seal fluid force and leakage under different journal offset

The Figure 5 shows the temporal change of the seal fluid force and the leakage under the offset 100μm and -100μm, which represents two distinguished state of the rotor system as described above. It is obvious to see that the seal fluid force under the offset of 100μm changes with time like the sine curve. While under the offset of -100μm, the sine curve of the seal fluid force are coupled with some half-frequency components. The similar change can also be seen in the leakage transit curve. This means the relative position between the journal and the seal can cause different seal fluid force components, which greatly affect the system stability and further affect the leakage. So the leakage changes in a similar way.

3.2. Transit characteristics

The Figure 6 shows the circumferential distribution of the total pressure at different position with different offsets. The distribution curves are similar to the sine curve at fixed positions marked in the small picture of the axis orbits. But the amplitude and the phase of the curves varies with different positions. This may have close relationship with the relative position between the journal and the labyrinth seal. When the journal offset is -100μm, the journal orbit has some half-frequency components as the figure shows. For the four points on the outer circle of the journal bearing locus, point C has the maximum peak-to-peak total pressure value while point A has the minimum value. This is probably because the point A is closer to the center of the seal, thus bringing a more even circumferential distribution of the total pressure. While for the point C, the clearance between the journal and the seal is the smallest, which makes the total pressure differs along the circumferential direction of the journal. Similarly, for the four points on the inner circle of the journal bearing locus, point B' is closer to the seal center and has a higher peak-to-peak total pressure value as well as a smaller clearance than point C'. When the journal offset is 100μm, the shape of the journal orbit is ellipse. With the change of the offset, point B'' has the smallest clearance and point D'' is closest to the center of the seal. As a result, the curve for the point D'' is more flat.

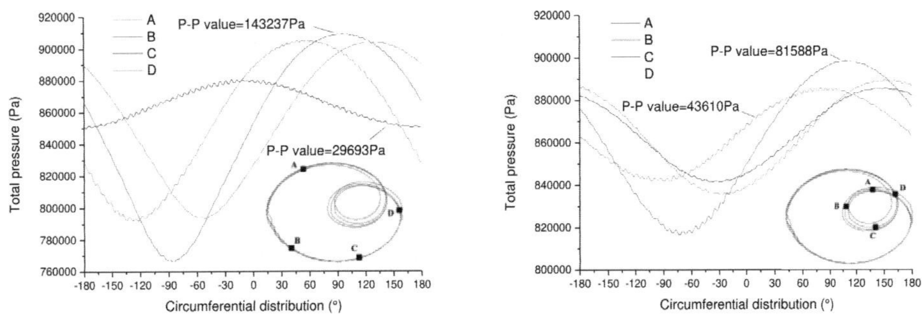

(a) Offset=-100μm

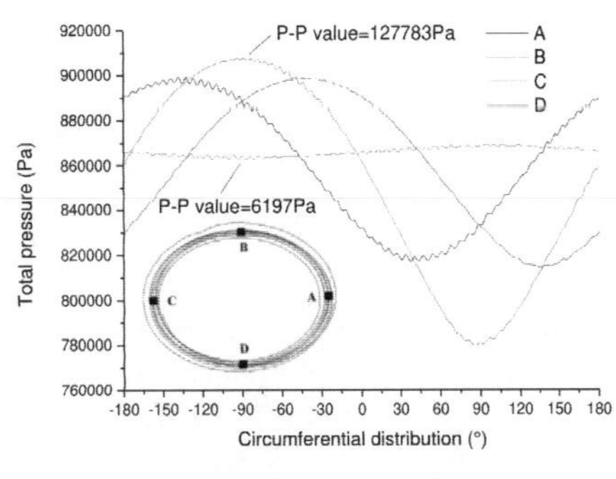

(b) Offset=100μm

Figure 6. The circumferential distribution of the total pressure at different position

The axial distribution of the total pressure at different position can be seen in Figure 7 From the curves, the pressure of the values decreased and are divided into several segments due to the effect of the labyrinth seal. Similar to the circumferential distribution of the total pressure, the total pressure distribution is different with the change of the journal position. For example, when the journal offset is 100μm, for the point B, the clearance between the journal and the seal is small, thus increasing the pressure drop of the seal.

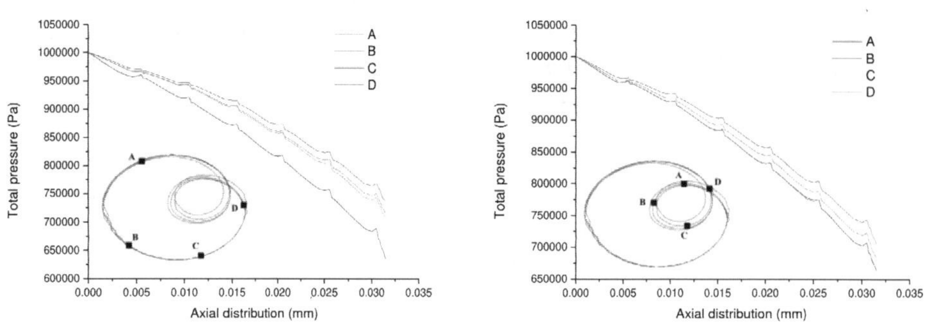

(a) Offset=-100μm

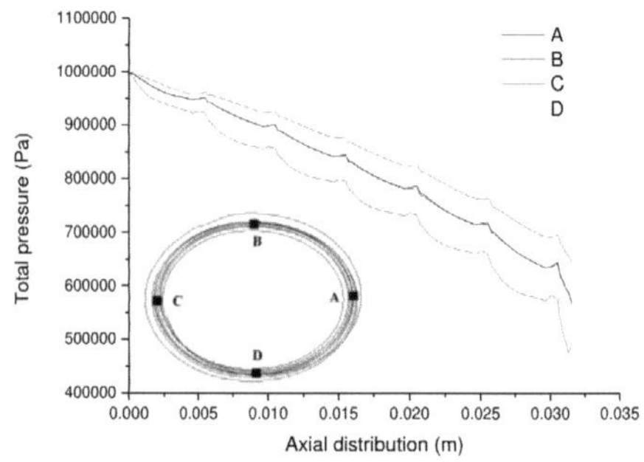

(b) Offset=100μm

Figure 7. The axial distribution of the total pressure at different position

From the above discussion, the pressure distribution changes regularly with the relative position between the journal and the seal. As a result, the nonlinear gas sealing force acted on the journal surface makes the journal move in a nonlinear orbit. In return, the nonlinear orbit will also result in the nonlinear change of the gas sealing force. The coupling effect will finally determine the stability of the system.

4. Experimental results

The Figure 8 shows the experimental vibration of a coupled rotor-seal system for the steam turbine. In the experiment, the labyrinth seal is designed to be allowed to adjust to a proper position. From the figure, when the time is around 70 min, the amplitude of the vibration decreases due to the adjustment of the labyrinth seal, which changes the eccentricity of the journal.

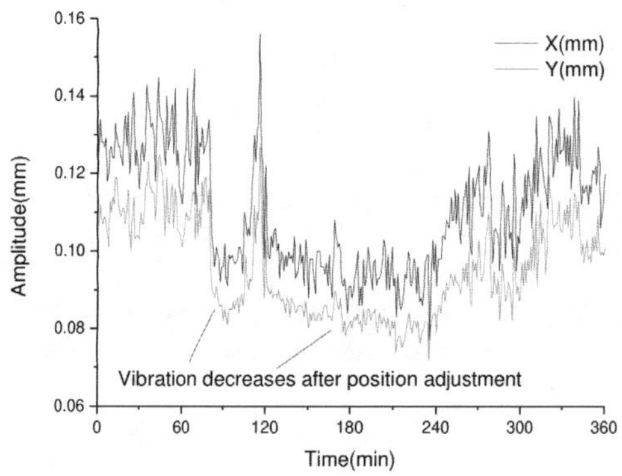

Figure 8. The experimental transit vibration amplitude

However, when the labyrinth seal position is adjusted to the original position at about 100 min, the vibration of the journal become severe. Similar phenomenon has been repeated at the time 120 min and 250 min. The experimental results have proved that the coupled rotor-seal system stability has close relationship with the relative position between the journal and the labyrinth seal and this conclusion agrees with the calculated results.

5. Conclusions

In this paper, a transit simulation method based on Fluent is introduced to model the coupled rotor-seal system. This method which involves a dynamic mesh method similar in journal bearing calculation can ensure fine mesh quality during the arbitrary motion of the journal. Experiment has been done to validate the conclusion. Some conclusions can be summed as followed:

1. The relative position between the journal and the labyrinth seal has close relationship with the gas sealing force, the leakage and the stability of the coupled rotor-seal system.
2. The circumferential distribution of the total pressure is similar to the sine curve. The amplitude of the curve increases with the eccentricity of the journal.

References

[1] Guinzburg, A., et al., The effect of inlet swirl on the rotordynamic shroud forces in a centrifugal pump. Journal of engineering for gas turbines and power, 1993. 115(2): p. **287-293**.

[2] Kirk, R.G., Evaluation of Aerodynamic Instability Mechanisms for Centrifugal Compressors—Part II: Advanced Analysis. Journal of Vibration and Acoustics, 1988. 110(2): p. **207-212**.

[3] Picardo, A. and D.W. Childs, Rotordynamic coefficients for a tooth-on-stator labyrinth seal at 70 bar supply pressures: measurements versus theory and comparisons to a hole-pattern stator seal. Journal of engineering for gas turbines and power, 2005. 127(4): p. **843-855**.

[4] Muszynska, A., Improvements in lightly loaded rotor/bearing and rotor/seal models. Journal of Vibration and Acoustics, 1988. 110(2): p. **129-136**.

[5] Muszynska, A. and D.E. Bently, Frequency-swept rotating input perturbation techniques and identification of the fluid force models in rotor/bearing/seal systems and fluid handling machines. Journal of Sound and Vibration, 1990. 143(1): p. **103-124**.

[6] Muszynska, A., Whirl and whip — rotor/bearing stability problems. Journal of Sound and vibration, 1986. 110(3): p. **443-462**.

[7] Li, S., Q. Xu and X. Zhang, Nonlinear dynamic behaviors of a rotor-labyrinth seal system. Nonlinear Dynamics, 2007. 47(4): p. **321-329**.

[8] Shen, X., et al., Experimental and numerical analysis of nonlinear dynamics of rotor – bearing – seal system. Nonlinear Dynamics, 2008. 53(1-2): p. **31-44**.

[9] Wang, W.Z., et al., A nonlinear model of flow-structure interaction between steam leakage through labyrinth seal and the whirling rotor. Journal of mechanical science and technology, 2009. 23(12): p. **3302-3315**.

[10] Childs, D.W. and J.K. Scharrer, Theory versus experiment for the rotordynamic coefficient of labyrinth gas seals: Part II—A comparison to experiment. Journal of Vibration and Acoustics, 1988. 110(3): p. **281-287**.

[11] Iwatsubo, T., Evaluation of instability forces of labyrinth seals in turbines or compressors. NASA. Lewis Res. Center Rotordyn. Instability Probl. in High-Performance Turbomachinery p **139-167**(SEE N 80-29706 20-37), 1980.

[12] Scharrer, J.K., Theory Versus Experiment for the Rotordynamic Coefficients of Labyrinth Gas Seals: Part I—A Two Control Volume Model. Journal of Vibration and Acoustics, 1988. 110(3): p. **270-280**.

[13] Arghir, M. and J. Fre^ne, Rotordynamic Coefficients of Circumferentially-Grooved Liquid Seals Using the Averaged Navier-Stokes Equations. Journal of tribology, 1997. 119(3): p. **556-567**.

[14] Arghir, M. and J. Fre^ne, Rotordynamic Coefficients of Circumferentially-Grooved Liquid Seals Using the Averaged Navier-Stokes Equations. Journal of tribology, 1997. 119(3): p. **556-567**.

[15] Hirano, T., Z. Guo and R.G. Kirk, Application of computational fluid dynamics analysis for rotating machinery—part ii: labyrinth seal analysis. Journal of engineering for gas turbines and power, 2005. 127(4): p. **820-826**.
[16] Ishii, E., et al., Prediction of Rotordynamic Forces in a Labyrinth Seal based on Three-Dimensional Turbulent Flow Computation. JSME International Journal Series C Mechanical Systems, Machine Elements and Manufacturing, 1997. 40(4): p. **743-748**.
[17] Li, Q., et al., A new method for studying the 3D transient flow of misaligned journal bearings in flexible rotor-bearing systems. Journal of Zhejiang University SCIENCE A, 2012. 13(4): p. **293-310**.

Interconnected State Control Method and Simulations of Four-corner Interconnected Air Suspension

Li Zhongxing[1], Zhou Ying[1], Zhang Xiao[2], Jiang Hong[2] and Xue Hongtao[1]

[1] School of Automotive and Traffic Engineering, Jiangsu University, 301 Xuefu Road, Zhenjiang, Jiangsu, China

[2] School of Mechanical Engineering, Jiangsu University, 301 Xuefu Road, Zhenjiang, Jiangsu, China

Email: zhxli@mail.ujs.edu.cn, 1652299795@qq.com, 1107332563@qq.com, 99998888@126.com, xueht@ujs.edu.cn

Abstract. The four-corner interconnected air suspension can effectively improve vehicles ride comfort, but the opening of the interconnection will intensify the roll or pitch of the car body, reducing the vehicle handling stability when the vehicles are in the turning, acceleration and deceleration conditions. To solve the above problems, an interconnected state control method is proposed for four-corner interconnected air suspension, the interconnection state of the pipeline is determined by the relationship between the torque produced by the air spring and the roll angles or pitch angles of the car body. The parameters of air spring pressure, car body roll angle and pitch angle are used in the control method to obtain the optimal interconnection state of four interconnected pipelines through control algorithm calculation. Finally, the opening and closing of four interconnected pipeline solenoid valves are controlled. In order to verify the effectiveness of the control method, the MATLAB/Simulink model of four-corner interconnected air suspension is established and simulated.

Key words: Air suspension, Four-corner interconnection, Simulation and modeling, Control method.

1. Introduction

The air suspension system has the characteristics of variable stiffness, low vibration frequency and convenient body control, which can effectively improve vehicle ride comfort, handling stability and road friendliness [1]. In order to further improve the performance of air suspension, many derivative structures have been proposed, especially interconnected air suspensions. According to different structures, the interconnected air suspension can be divided into lateral interconnected air suspension, longitudinal interconnected air suspension and four-corner interconnected air suspension etc. Four-corner interconnection can be regarded as the combination of lateral and longitudinal interconnection air suspension namely the adjacent air springs of air suspension are connected, as shown in Fig.1[2]. In terms of performance, four-corner interconnected air suspension has the characteristics of both lateral and longitudinal interconnected air suspension. Compared with the traditional air suspension, the four-corner interconnected air suspension can improve the ride comfort of vehicles on rough roads, reduce torsional load of the vehicle body and improve the vibration isolation performance of vehicles [3]. However, the lateral interconnection will aggravate the roll of the vehicle when the vehicle is in a turning condition [4]. The longitudinal interconnection will aggravate the pitch of the vehicle when the vehicle is in acceleration and deceleration [5].

Recently, some research on interconnect ion air suspensions have been reported. [6-7] established a nonlinear model of the longitudinal-connected air suspension of tri-axle semi-trailer. The effects of driving conditions and suspension parameters on the dynamic load distribution of the longitudinal-connected air suspension of the tri-axle semi-trailer were studied. [8-9] verified the relationship between increasing the diameter of longitudinal interconnected pipes and improving the dynamic load distribution through t-test. It was concluded that the dynamic load between suspension and chassis of heavy vehicles can be effectively reduced by using larger longitudinal interconnected pipes than the standard size. [10] proposed the imitated skyhook control theory of laterally interconnected air suspension. The theory took the vehicle body roll angle and the "mass roll angle under the spring" as the monitoring parameters and used their product to judge the opening and closing of the air suspension interconnection state.

However, there are few studies concentrated on the interconnection state control of four-corner interconnected air suspension. In this paper, a control method for the interconnected state of four-corner interconnected air suspensions is presented. Firstly, the interconnection state of pipeline can be determined according to the relationship between the air spring torque and the roll angles or pitch angles of vehicle body. Secondly, the method takes the vehicle body roll angle, pitch angle and the air pressure of each air spring as the monitoring parameters. Finally, the optimum interconnection state of the air suspension is obtained through MATLAB/Simulink simulations.

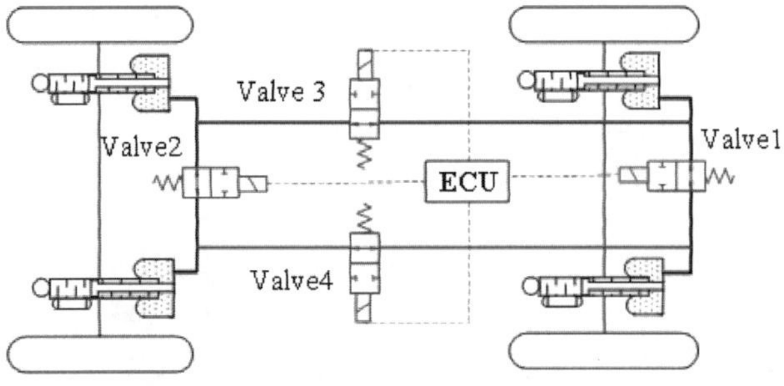

Figure 1. Four-corner interconnected air suspension

2. The establishment of vehicle vibration models

In order to study effectively the roll, pitch and vibration isolation characteristics of four-corner interconnected air suspension on uneven road surface, a seven-degree-of-freedom vehicle dynamic model was established. The suspension dynamic model is shown in Fig.2.

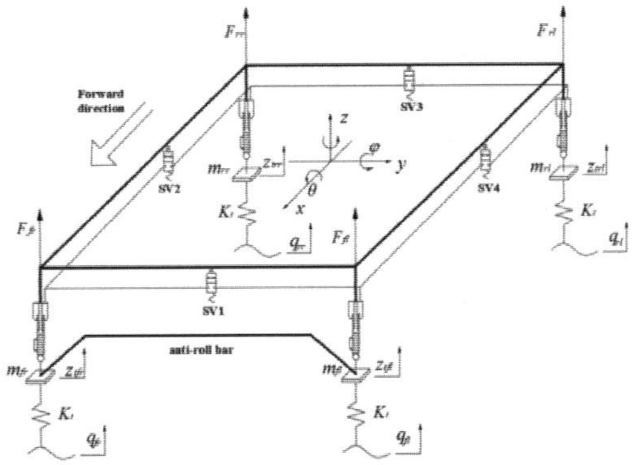

Figure 2. Full vehicle dynamic model of four-corner interconnected air suspension

$$
\begin{cases}
m_{fl}\ddot{z}_{tfl} = K_t(q_{fl} - z_{tfl}) - F_{fl} + M_{bf}g/2 - k_{bar}\theta/B_f \\
m_{fr}\ddot{z}_{tfr} = K_t(q_{tfr} - z_{tfr}) - F_{fr} + M_{bf}g/2 + k_{bar}\theta/B_f \\
m_{rl}\ddot{z}_{trl} = K_t(q_{rl} - z_{trl}) - F_{rl} + M_{bf}g/2 \\
m_{rr}\ddot{z}_{trr} = K_t(q_{rr} - z_{trr}) - F_{rr} + M_{bf}g/2 \\
M_b\ddot{z}_b = F_{fl} + F_{fr} + F_{rl} + F_{rr} - M_b g \\
I_r\ddot{\theta} = (F_{fl} - F_{fr})B_f/2 + (F_{rl} - F_{rr})B_r/2 + M_b g H_r\theta - k_{bar}\theta + M_b a_y H_r \\
I_p\ddot{\varphi} = (F_{rl} + F_{rr})L_r - (F_{fl} + F_{fr})L_f + M_b g H_p\varphi
\end{cases}
\quad (1)
$$

where, m_{fl} 、 m_{fr} 、 m_{rl}、 m_{rr} are front left, front right, rear left, rear right unsprung masses respectively; z_{tfl} 、 z_{tfr} 、 z_{trl}、 z_{trr} are front left, front right, rear left, rear right displacements of unsprung masses respectively; K_t is vertical stiffness of tire. q_{fl} 、 q_{fr} 、 q_{rl}、 q_{rr} are front left, front right, rear left, rear right vertical displacements of the road respectively; F_{fl} 、 F_{fr} 、 F_{rl}、 F_{rr} are suspension forces of front left, front right, rear left, rear right wheels respectively. The suspension force of a wheel is the resultant force of the air spring force and damping force at the wheel. M_{bf} and M_{br} are body mass carried by the front and rear axle of the vehicle, k_{bar} is the roll stiffness of anti-roll bar; B_f and B_r are respectively front and rear wheel treads; M_b is the body mass, z_b is vertical displacement of spring mass center of mass, I_r is the rotary inertia of the body around the roll axis (X axis), θ is the body roll angle and the right deviation of the car body is positive, I_p is the rotary inertia of the body around the pitch axis (Y axis), φ is the body pitch angle, which is assumed positive when the body bends forward, H_r is the vertical distance between the center of sprung mass and the roll axis, H_p is the vertical distance between the center of sprung mass and the pitch axis, a_y is lateral acceleration of vehicle body, L_f and L_r are respectively distance from front and rear axle to the centroid.

The air spring is regarded as a variable mass thermal insulation system according to the first law of system thermodynamics [11]:

$$P\left(\frac{V}{m}\right)^k = \text{const} \quad (2)$$

where, P is the absolute pressure of air in the air spring, V is the air spring chamber volume, m is the air mass in the air spring, k is adiabatic exponent and its value is 1.4 for air.

The pressure response in the front left air spring can be obtained from the Eq. (3):

$$\begin{cases} P = P_0\left(\dfrac{mV_0}{m_0V}\right)^k \\ V = V_0 - \dfrac{dV}{h}\cdot f_d \end{cases} \tag{3}$$

where, m_0 is the initial gas mass of the air spring, V_0 is the initial chamber volume of the air spring, dV/h is the volume change rate of air springs, f_d is the suspension working space, namely the height change of air spring.

The spring force of the air spring can be calculated by the relationship between the pressure and the effective area of the air spring:

$$Fa = (P - P_a)A_e \tag{4}$$

where, F_a is air spring force, P_a is atmospheric pressure, A_e is the effective area of the air spring.

The adjacent air springs are connected through pipelines, when the pressure difference at both ends of the pipeline produces gas flow, the flow characteristics of the pipeline can be equivalent to the characteristics of the throttle orifice. The mass flow rate of gas flowing through the pipeline can be expressed as follows [12]:

$$\dot{m}(0,t) = \begin{cases} AP_{up}\sqrt{\dfrac{1}{RT_{up}}\dfrac{2k}{k-1}\left[\left(\dfrac{P_{down}}{P_{up}}\right)^{\frac{2}{k}} - \left(\dfrac{P_{down}}{P_{up}}\right)^{\frac{k+1}{k}}\right]}\operatorname{sgn}(P_{down}-P_{up}) & \dfrac{P_{down}}{P_{up}} > v_{cr} \\ AP_{up}\left(\dfrac{2}{k+1}\right)^{\frac{1}{k-1}}\sqrt{\dfrac{1}{RT_{up}}\dfrac{2k}{k+1}}\operatorname{sgn}(P_{down}-P_{up}) & \dfrac{P_{down}}{P_{up}} \le v_{cr} \end{cases} \tag{5}$$

where, A is the effective area of orifice, P_{up} is absolute upstream gas pressure, P_{down} is absolute downstream gas pressure, T_{up} is upstream gas temperature, R is the ideal gas constant.

The parameters of simulation model under Matlab/Simulink environment are displayed in Tab.1.

Table 1. Full car model parameters.

Parameter	Value	Parameter	Value
Body mass M_b [kg]	6700	Tire stiffness K_t [kN/m]	210
Wheel mass M_t (front/rear) [kg]	300/500	Distance between shafts L [m]	3.9
X axis rotary inertia I_r [kg·m^2]	2857	Front axle wheelbase B_f [m]	1.7
Y axis rotary inertia I_p [kg·m^2]	4983	Rear axle wheelbase B_r [m]	1.5
Sprung mass centroid height H[m]	1.1	Anti-roll bar roll stiffness k_{bar} [N·m/°]	4433
Distance from centroid to front axle L_f [m]	2.7	Distance from centroid to rear axle L_r [m]	1.3

3. The principle of interconnection state control method

The four-corner interconnected air suspension can improve vehicle ride comfort under road excitation, but when the vehicle is in the turning, acceleration and deceleration conditions, it will aggravate the roll or pitch of the body, and reduce the vehicle handling stability. The front axle lateral interconnection is taken as an example, as shown in Fig.3.

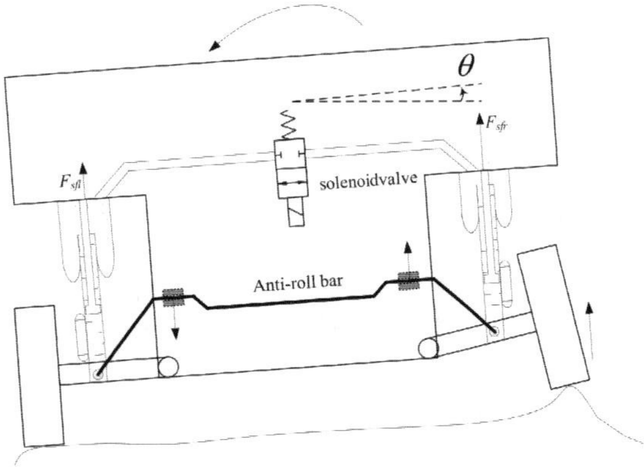

Figure 3. Road surface excitation diagram for front axle laterally interconnected air suspension

When the front right wheel is subjected to road excitation, the air pressure in the left and right air springs of the front axle is P_{fl} and P_{fr} respectively, the effective bearing area is S_{fl} and S_{fr} respectively. If the interconnection is closed, the air spring on the right side is compressed, and the air spring pressure P_{fr} increases. The spring force difference between the left and right suspensions is different. The spring force difference between the left and right suspension can be calculated as follows:

$$\Delta F = F_{sfr} - F_{sfl} = P_{fr}S_{fr} - P_{fl}S_{fl} \tag{6}$$

where, F_{sfl}、F_{sfr} are spring forces of left and right air springs on front axle. M_f is body roll torque subjected to the front axle, it can be calculated as follows:

$$M_f = F_{sfr} \cdot T_j / 2 - F_{sfl} \cdot T_j / 2 = \Delta F \cdot T_j / 2 \tag{7}$$

Resulting from the great difference between F_{sfl} and F_{sfr} of air spring force, the car body is subjected to bigger roll torque, which results in bigger roll acceleration, bigger body roll angle and smaller vehicle handling stability. If the interconnection is opened, the air of the compressed air spring on the right flows to the air spring on the left side, and the air pressure of the compressed air spring tends to be balanced due to the flow of the air. Both of the air pressures of the two air springs after balancing are P_f, the difference between the left and right spring forces is shown as follows:

$$\Delta F = F_{sfr} - F_{sfl} = P_f \left(S_{fr} - S_{fl} \right) \tag{8}$$

The effective bearing area S_{fl} and S_{fr} of the left and right air springs vary little during their working stroke, therefore, when the air spring is interconnected, the spring force difference ΔF decreases significantly. Simultaneously, the roll torque M_f and the body roll angle are reduced correspondingly.

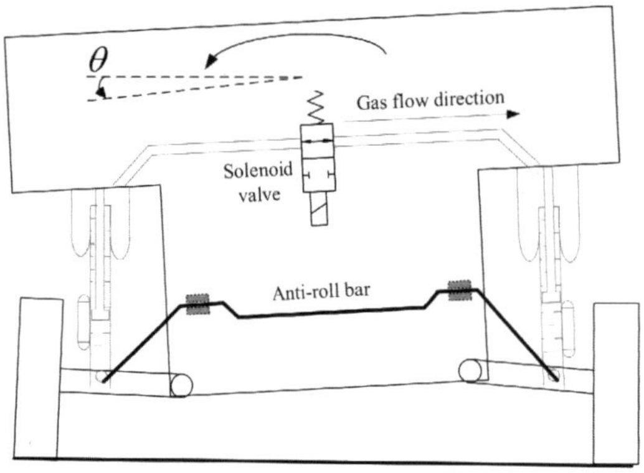

Figure 4. Turning diagram of horizontal road surface for front axle laterally interconnected air suspension

In the turning condition, the front axle lateral interconnection is displayed in Fig.4.When the vehicle turns to right on the horizontal road surface, axial load transfer is generated by centrifugal force, the load on the left air spring increases, and the load on the right air spring decreases. Therefore, the air pressure on the left air spring rises, and the air pressure on the right side of the air spring decreases. If the front axle lateral interconnection opens at this time, the compressed air on the left side flows to the right under pressure, further aggravating the body roll. If the interconnection is closed, the air pressure on the left air spring increases, and the air pressure on the right side of the air spring decreases.

Therefore, some conclusions can be drawn: when the direction of roll torque produced by air spring is the same as the direction of body roll, the opening of interconnection can reduce the roll torque produced by air spring to promote body roll; When the direction of the roll torque produced by the air spring is opposite to the body roll direction, the interconnection closure can increase the anti-roll torque produced by the air spring and restrain the further roll of the body.

The air pressure of the air spring is closely related to the change of the interconnection state. The body roll angle θ is an important indicator of body roll, its symbols can characterize the direction of roll, and its absolute value can characterize the degree of roll. The former axis lateral interconnection is taken as an example, when the product of the air pressure difference between the left and right air springs and the roll angle is greater than 0, the roll torque provided by the front axle air spring is the same as the roll direction of the car body, opening lateral interconnection; when the product of the air pressure difference between the left and right air springs and the roll angle is less than 0, that is, the roll torque provided by the front axle air spring is opposite to the roll direction of the car body, closing the lateral interconnection. It can be expressed as follows:

$$\begin{cases} \left(P_{fl} - P_{fr}\right)\theta > 0 & Qvalue = 1 \\ \left(P_{fl} - P_{fr}\right)\theta \le 0 & Qvalue = 0 \end{cases} \tag{9}$$

where, *Qvalue* is the interconnection state of the front axle lateral connection, when the value is 1, the interconnection opens; when the value is 0, the interconnection is closed. P_{fl} and P_{fr} are the air springs pressure on the left and right sides of the front axle.

Similarly, the interconnection state of the rear axle air spring can be judged as follows:

$$\begin{cases} \left(P_{rl} - P_{rr}\right)\theta > 0 & Hvalue = 1 \\ \left(P_{rl} - P_{rr}\right)\theta \leq 0 & Hvalue = 0 \end{cases} \tag{10}$$

where, *Hvalue* is the interconnection state of the rear axle lateral connection, when the value is 1, the interconnection opens; when the value is 0, the interconnection is closed. P_{rl} and P_{rr} are the air springs pressure on the left and right sides of the rear axle.

In addition to two lateral interconnected pipelines, there are two longitudinal interconnected pipelines in the four-corner interconnected air suspension. According to the above rules, when judging the interconnected state of the longitudinal interconnected pipelines, the air pressure difference between the front and rear air springs on the same side and the pitch angle of the car body should be used. The interconnection state of the longitudinal interconnected pipelines can be judged by follows:

$$\begin{cases} \left(P_{fl} - P_{rl}\right)\varphi > 0 & Zvalue = 1 \\ \left(P_{fl} - P_{rl}\right)\varphi \leq 0 & Zvalue = 0 \end{cases} \tag{11}$$

$$\begin{cases} \left(P_{fr} - P_{rr}\right)\varphi > 0 & Yvalue = 1 \\ \left(P_{fr} - P_{rr}\right)\varphi \leq 0 & Yvalue = 0 \end{cases} \tag{12}$$

where, φ is the pitch angle of the car body; *Hvalue* and *Yvalue* respectively represent the left and right longitudinal interconnection state. When the value is 1, the interconnection opens; when the value is 0, the interconnection is closed.

4. The comparison of simulation results

In order to verify the vibration isolation characteristics of vehicles, sine sweep excitation with amplitude of 0.015m is applied to four wheels of vehicle, the excitation frequency is 1Hz~20Hz, increasing 0.25Hz per second in the simulation model. The excitation frequencies of each wheel are always the same, but the phases of each wheel are different. In order to apply both the lateral excitation and the pitch excitation, the phase difference between the front and rear, left and right are both 180°. Therefore, the wheel phases of front left, front right, rear left and rear right are 0°, 180°, 180°and 0°respectively. The frequency domain response of mass acceleration on the front left spring of the air suspension under four interconnected states and control methods are shown in Fig. 5.

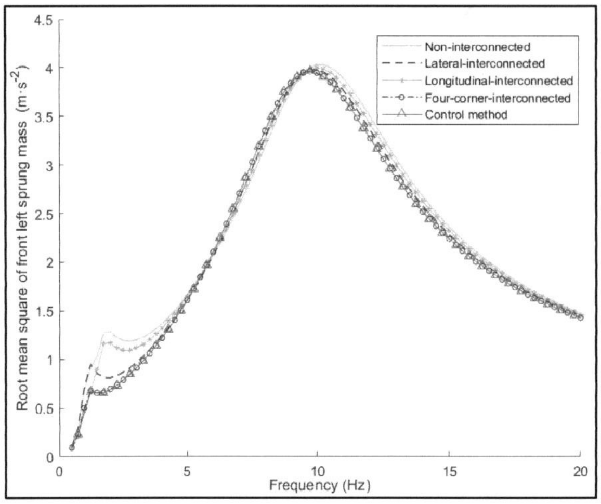

Figure 5. The response of front left sprung mass with different interconnected modes in frequency domain

When the phase difference of the sinusoidal sweep excitation between the front and back as well as the left and right is both π, the resonance peak of mass under the spring will appear at 1-2Hz, and the resonance peak of the mass on the front left spring will appear at 9.8-10.2Hz. The root mean square (RMS) of mass acceleration on the front left spring under the interconnection control method is essentially same as that on the front left spring under the four-corner interconnection condition. The RMS resonance peak value of mass acceleration on the front left spring under four-corner interconnection control method is lowest compared with the other three interconnected states, and achieves the desired control effect.

In order to verify the control effect on the random road surface, the driving conditions are set: D level road, speed 30km/h, straight line driving. To express clearly, simulation time 10s. Vehicle body roll angle, pitch angle and root mean square acceleration of front left sprung mass under several interconnection states and interconnection control method are compared in Figures 6, 7 and 8.

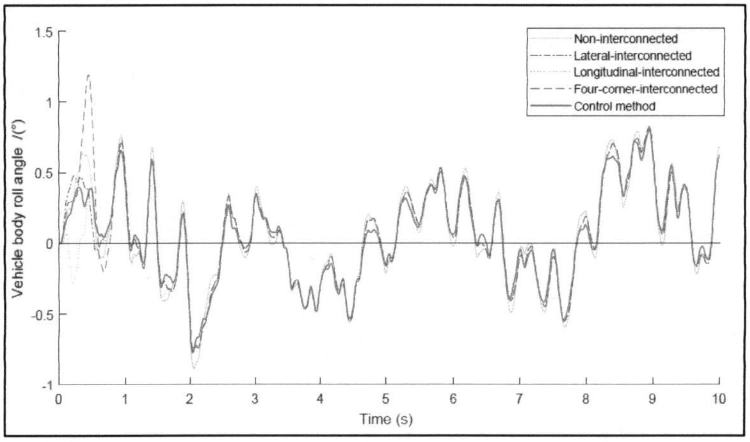

Figure 6. The vehicle body roll angle in different interconnected modes on the random road surface

The root mean square values of the body roll angles under the control of non-interconnection, lateral-interconnection, longitudinal interconnection, four-corner-interconnection and interconnection method are at 0.3548°, 0.3378°, 0.3434°,0.3609° and 0.3221° respectively. Compared with non-interconnection, lateral-interconnection, longitudinal interconnection and four-corner-interconnection, the root mean square values of the body roll angles under control method decreases by 9.2%、4.6%、2.1% and 3.88% respectively.

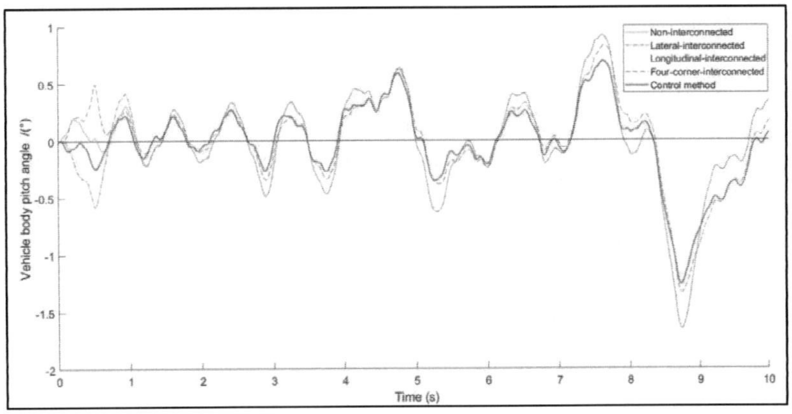

Figure 7. The vehicle body pitch angle in different interconnected modes on the random road surface

The vehicle body pitch angle under interconnected control method has been at a lowest level than other interconnection modes. The root mean square values of the body pitch angles under the control of non-interconnection, lateral-interconnection, longitudinal interconnection, four-corner-interconnection and interconnection method are at 0.4244°, 0.4337°, 0.3814°,0.3771° and 0.3405° respectively. Compared with non-interconnection, lateral-interconnection, longitudinal interconnection and four-corner-interconnection, the root mean square values of the body pitch angles under control method decreases by 19.8%、21.5%、10.7% and 9.7% respectively.

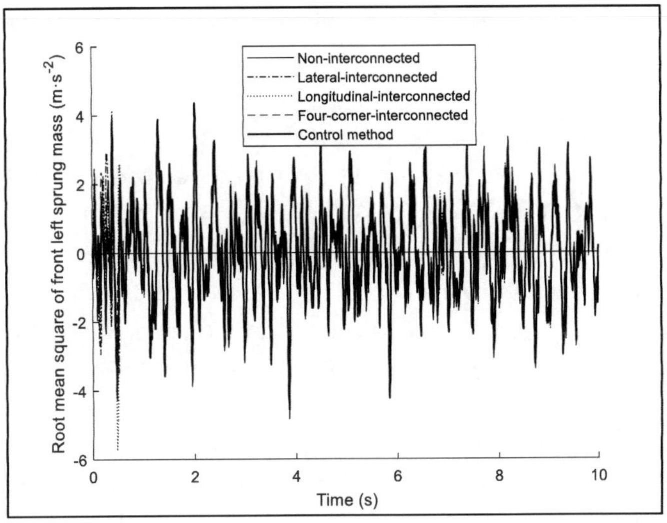

Figure 8. Root mean square of mass acceleration on front left spring in different interconnected modes on the random road surface

The mean square values of mass acceleration on front left spring under interconnected control method have been at a lowest level compared with other interconnection modes. The root mean square values of mass acceleration on front left spring under the control of non-interconnection, lateral-interconnection, longitudinal interconnection, four-corner-interconnection and interconnection method are at 1.3950m/s^2, 1.3556m/s^2, 1.4032m/s^2, 1.3497m/s^2 and 1.3240m/s^2 respectively. Compared with non-interconnection, lateral-interconnection, longitudinal interconnection and four-corner-interconnection, the mean square values of mass acceleration on front left spring under interconnected control method decreases by 5.0%、2.3%、5.6% and 1.9% respectively.

5. Conclusions

The influence of air spring torque of interconnected air suspension on vehicle body roll or pitch angles due to the change of interconnection state is analyzed by establishing the vehicle models. Based on it, the principles of interconnection state control method are formulated. Finally, some results from Matlab/Simulink simulation can give the conclusion as follows:

(1)A sinusoidal scanning excitation of 1 Hz to 20 Hz is applied to the vehicle body in the Simulink environment. The RMS resonance peak value of mass acceleration on the front left spring under four-corner interconnection control method is lowest.

(2) In the random road simulation environment, the vehicle body roll angle, pitch angle and the root mean square value of mass acceleration on the front left spring under interconnection method are 0.3221°, 0.3405°and 1.3240m/s^2 respectively. All are the smallest compared with other interconnection modes.

The results show that this control method can not only reduce the roll angle and pitch angle caused by road excitation, but also improve the anti-roll and anti-pitch ability of the air suspension under turning, acceleration and deceleration conditions, and maintain good vibration isolation characteristics.

Acknowledgements

This project is supported by the National Natural Science Foundation of China (Grant No.51575241).

References

[1] Ma, X., Wong, P. K., Zhao, J., Zhong, J. H., Ying, H. and Xu, X. "Design and testing of a nonlinear model predictive controller for ride height control of automotive semi-active air suspension systems", IEEE Access, Vol.6, 2018, p.63777-63793.

[2] Friedrich, W. M., Mathias, S. and Michael, F. "Interconnected air suspension systems: the influence on ride comfort in testing and simulation", ATZ Auto technology, Vol.9, 2009, p.58-61.

[3] Jiang, H., Qian, K., Zhu, C. and Sun, Y. "Characteristics of Ride Comfort and Torsion Elimination of Four-Corner Interconnected Air Suspension System", Journal of Southwest Jiaotong University, Vol. 6, 2015, p.19.

[4] Li, Z., Cui, Z., Xu, X., and Qiu, Y. "Experimental study on dynamic characteristics of interconnected air suspension", Science and Technology and Engineering, Vol.14(14) 2014, p.82-86.

[5] Kat, C. J. and Schalk Els, P. "Interconnected air spring model", Mathematical and Computer Modelling of Dynamical Systems", Vol.15(4), 2009, 353-370.

[6] Chen, Y., He, J., King, M. J., Chen, W., Wang, C. and Zhang, W. "Model development and dynamic load-sharing analysis of longitudinal-connected air suspensions", Strojniški vestnik–Journal of Mechanical Engineering, Vol.59(1), 2013, p14-24.

[7] Chen, Y., He, J., King, M., Chen, W. and Zhang, W. "Effect of driving conditions and suspension parameters on dynamic load-sharing of longitudinal-connected air suspensions", Science China Technological Sciences, Vol.56(3) ,2013, p666-676.

[8] Davis, L. E. and Bunker, J. M. "Altering heavy vehicle air suspension dynamic forces by modifying air lines", International Journal of Heavy Vehicle Systems (IJHVS) I, Vol.18(1) ,2011, 1-17.

[9] Davis, L. and Bunker, J. M. "Dynamic load sharing in heavy vehicle suspensions-further analysis of the equalisation effects of larger air lines", Modern Traffic and Transportation Engineering Research, Vol. 3(1), 2014, p.1-6.

[10] Li, Z., Ju, L., Jiang, H. and Xu, X. "Imitated skyhook control of a vehicle laterally interconnected air suspension", International Journal of Vehicle Design, Vol.74(3), 2017, p.204-230.

[11] Cai Maolin, "Theory and practice of Modern Pneumatic Technology Second talk: Charging and discharging of fixed volume chambers", Hydraulics Pneumatics & Seals, Vol.3,2007, p.43-47.

[12] Li Mei. "Mechanism Analysis on Dynamic Characteristics of Air Spring System with Auxiliary Chamber", Jiangsu University, 2012.

Study on the ultimate load carrying capacity of journal bearings with different materials and surface treating processes

Hulin Li[1,*], Yanzhen Wang[2] and Zhongwei Yin[3]

[1, 3] School of design, Shanghai Jiao Tong University, Shanghai, China
[2] CSSC 708 Research Institute, Shanghai, China

Email:gslhl@126.com; merinawang@163.com; yinzw@163.com

Abstract. The ultimate load carrying capacity of journal bearings with different materials and surface treatment processes was studied in the present work. The load carrying capacity and backside temperature of bearing were examined using a bearing fatigue test machine (Glacier Vandervel, made in UK). The results showed that the copper-based sputtering bearing has the highest ultimate bearing capacity and the fatigue strength reaches 131.3 MPa. The bearing capacity of copper-based electroplated bearings is 99.7 MPa, which is lower than that of copper-based sputtering bearings. The bearing capacity of high tin aluminum alloy is obviously higher than that of white alloy bearing, and the bearing capacity of the two bearings is 72.7 MPa and 36.3 MPa, respectively. Finally, the mechanisms of fatigue failure of the bearings were analyzed.

Keywords: ultimate load carrying capacity, sputtering bearing, electroplating bearing, Aluminium-based alloy bearing, tin-based alloy bearing

1. Introduction

As one of the most important component of diesel engines, oil-lubricated journal bearings are designed to convert the reciprocating motion of pistons into the rotary motion of the crankshaft[1]. It is critical for the stable operation of diesel engines that the bearing material has excellent load carrying and tribological behavior. Bearing material is generally required to have good embedability, conformability, seizure resistance, and most importantly excellent tribological performance under oil-lubricated conditions [2-4].

The load carrying capacity and tribological behavior of journal bearings materials have been investigated by many related studies. Bican and Savaskan [5] studied the wear behaviors of Al–25Zn–

3Cu–3Si alloy using a block-on-disc machine and compared with the performance of 60/40 brass and SAE 65 bronze, the results showed that the Al–25Zn–3Cu–3Si alloy showed the best wear resistance. Kumar et al [6] studied the friction and wear characteristics of Cu-Sn alloy reinforced with molybdenum disulfide (MoS2) fabricated by a powder metallurgy route. The results showed that the wear rate of composites is the lowest when MoS2 content is 7 wt%. Miguel et al [7] studied the effect of spraying process on tribological behavior of bronze–alumina composite coatings by ball on disc dry sliding tests, and the results showed that the wear rate of composites was significantly reduced by Al2O3 particles.

The load carrying capacity and tribological behaviors between different bearing materials vary greatly. Until now, some research studied the load carrying capacity and frictional behavior of journal bearing under static load, but the performances under actual dynamic load are unclear. Thus, this paper focuses on the bearing capacity and operating characteristics of four kinds of bearing (sputtering AlSn20Cu bearing, electroplating AlSn20Cu bearing, AlSn40Cu alloy bearing and ZSnSb8Cu4 alloy bearing) under dynamic load to provide technical support for the practical application of these kinds of bearings.

2. Experimental section

2.1 Materials and specimen preparation

In the experiment, the ultimate load carrying capacity of four kinds of bearing prepared by different matrix materials and surface treating processes were tested. The shaft was made of 42CrMo steel and the bearings were made of bimetal composite or three-layer metal composite. The copper-based sputtering bearing is composed of 10# steel back, CuPb22Sn2 alloy substrate layer and AlSn20Cu alloy sputtering layer, which is abbreviated as CBAS-Bearing in this paper. The copper-based electroplating bearing is composed of 10# steel back, CuPb22Sn2 alloy substrate layer and PbSn18Cu2 alloy electroplating layer, which is abbreviated as CBPE-Bearing in this paper. The high tin aluminum alloy bearing is composed of 10# steel back, AlSn40Cu alloy substrate layer, which is abbreviated as HTAB-Bearing in this paper. The tin-based bearing is composed of 10# steel back, ZSnSb8Cu4 alloy substrate layer, which is abbreviated as TB-Bearing in this paper.

2.2. Ultimate load carrying capacity test

Figure 1. The schematics of the test assembly

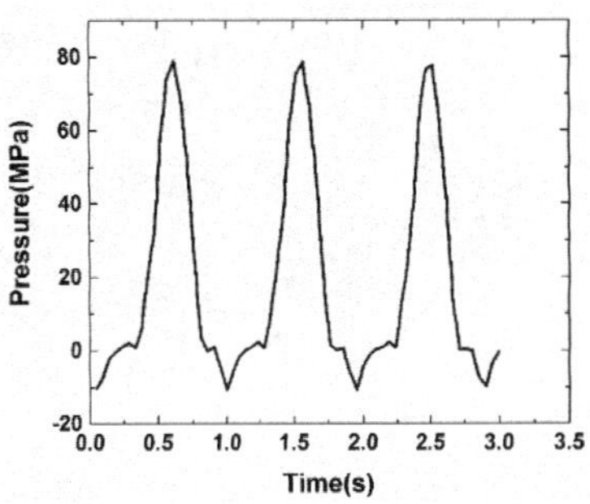

Figure 2. Test load curve

Ultimate load carrying capacity tests of bearings were performed using a bearing fatigue strength test machine (Glacier Vandervel, made in UK), as shown in Fig. 1. Each test runs for 20 hours, and then checks the damage of bearing. If the bearing passes the test (no damage visible to the naked eyes), add one level load (7.0 MPa) to continue the test. Repeat the above operation until the bearing is fatigue damaged or wear out. The previous level load when the bearing is damaged is determined as the ultimate load carrying capacity of the bearing. The test shaft diameter is $52.7^{-0.030}_{-0.013}$mm, bearing outer diameter is 56.426mm, bearing thickness is $1.825^{\ 0}_{-0.006}$mm, and bearing width is 29.5 mm. For the ultimate load carrying capacity tests, the rotational speed is 3000 r/min and the running time is 20 hours (3.6×10^6 cycles) for each test. As shown in Fig. 2, the loading frequency is 50 Hz and sampling frequency is 1000 Hz.

3. Results and discussion

Fig. 3 shows the variation of the backside temperature of bearings with load and time. As shown in Fig.3, the backside temperature of bearings for different material and surface treatment processes increases first and then remains basically unchanged. In addition, for all the bearings the backside temperature increases with the increase of load. As can be seen from Fig. 3a and Fig. 3b, under the almost same load conditions, the backside temperature of bearings with electroplating layer is significantly lower than that of bearings with sputtering layer. The possible reasons are that the hardness and the friction coefficient of PbSn18Cu2 electroplated coating are lower than that of AlSn20Cu sputtering coating.

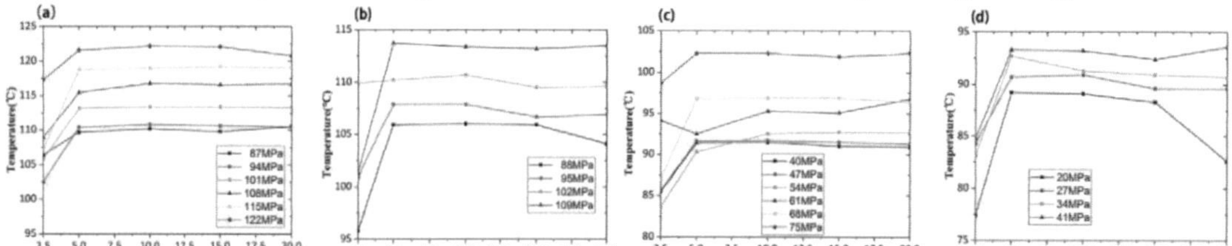

Figure 3. Backside temperature of bearings vs. load and time (a: CBAS-Bearing; b: CBPE-Bearing; c: HTAB-Bearing; d: TB-Bearing)

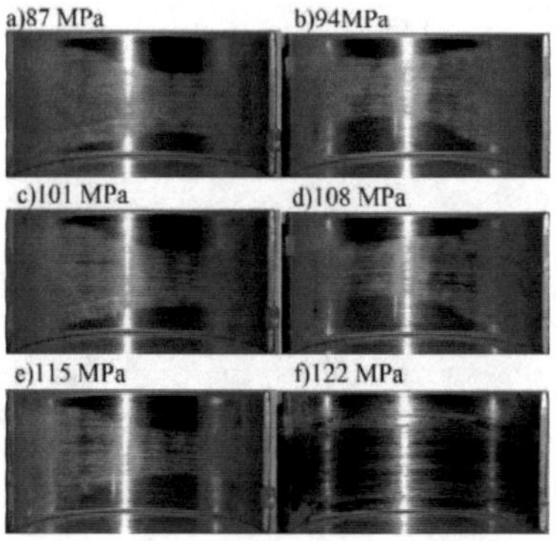

Figure 4. Worn surface of CBAS-Bearing

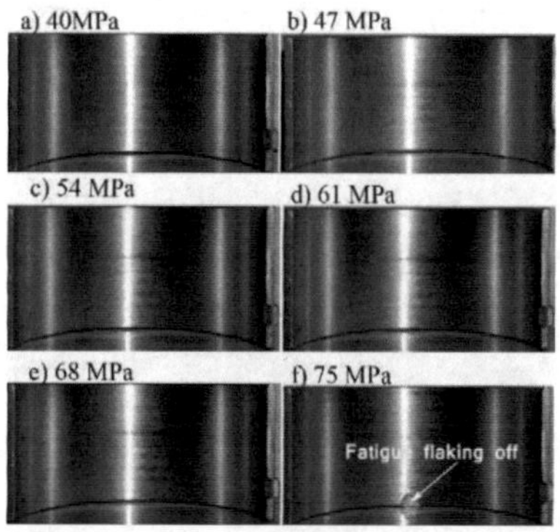

Figure 5. Worn surface of HTAB-Bearing

Fig. 4 shows the digital camera photo of worn surface of CBAS-Bearing under different loads. The results showed that, as the increase of load, the wear zone develops from the end surfaces of the bearing toward the middle plane of the bearing and the wear damage increases gradually. The AlSn20Cu sputtering layer on the bearing surface is completely worn off when the load is increased to 122 MPa.

IOP Conf. Series: Materials Science and Engineering **689** (2019) 012014 doi:10.1088/1757-899X/689/1/012014

However, the CuPb22Sn2 alloy substrate layer of CBAS-Bearing is still in the good condition. Therefore, the main failure mode of CBAS-Bearing is the abrasion wear of sputtered coating.

Fig. 5 shows the digital camera photo of worn surface of HTAB-Bearing under different loads. The results showed that the wear of AlSn40Cu alloy bearing is not significant when the load increases to 68 MPa. The main reason is that the lubrication film between the bearing and the shaft is thick when the load is small, thereby avoiding the direct contact of the bearing and shaft. However, when the load increases to 75 MPa, fatigue flaking off as shown in Fig. 5f appears on the worn surface of AlSn40Cu alloy layer. Therefore, the failure of the HTAB-Bearing is mainly due to the fatigue failure caused by the test load approaching the material's fatigue limit.

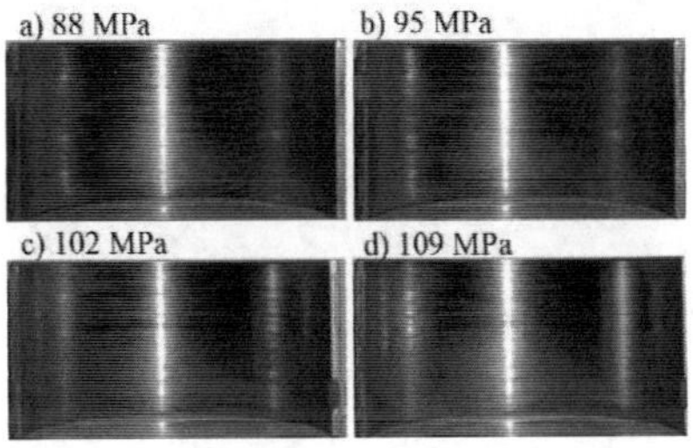

Figure 6. Worn surface of CBPE-Bearing

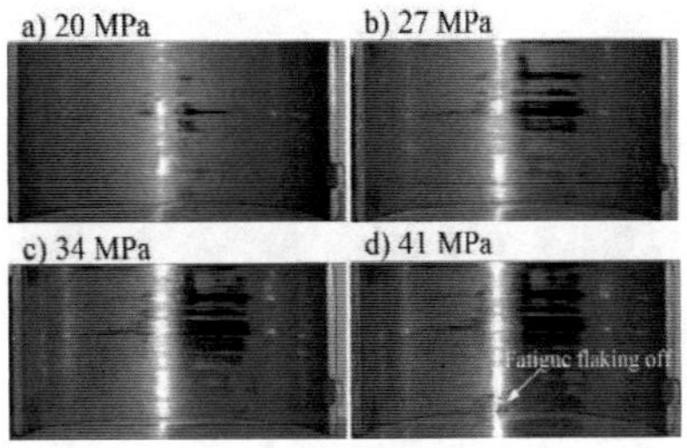

Figure 7. Worn surface of TB-Bearing

Fig. 6 shows the digital camera photo of worn surface of CBPE-Bearing under different loads. The results showed that the wear of PbSn18Cu2 electroplated coating aggravates as the increase of load. The PbSn18Cu2 electroplated layer is completely worn through when the load increases to 109 MPa. Like the CBAS-Bearing, the CuPb22Sn2 alloy substrate layer of CBPE-Bearing is also intact, which indicates that the main failure mode of CBPE-Bearing, like that of CBAS-Bearing, is the abrasion wear of the coating.

Fig. 7 shows the digital camera photo of worn surface of TB-Bearing under different loads. The results showed that the abrasion wear of ZSnSb8Cu4 alloy substrate layer increases with the increase of load. As we all know, the hardness and wear resistance of TB-Bearing are much worse than those of CBAS-Bearing, CBPE-Bearing and HTAB-Bearing. The wear of TB-Bearing is not serious because of the small load and the thick oil film between bearing and shaft. However, it can be seen from the Fig.3d that when the test load is increased to 41MPa, the fatigue is happened. Therefore, same as the HTAB-Bearing, the main failure mode of TB-Bearing is fatigue damage of based alloy layer.

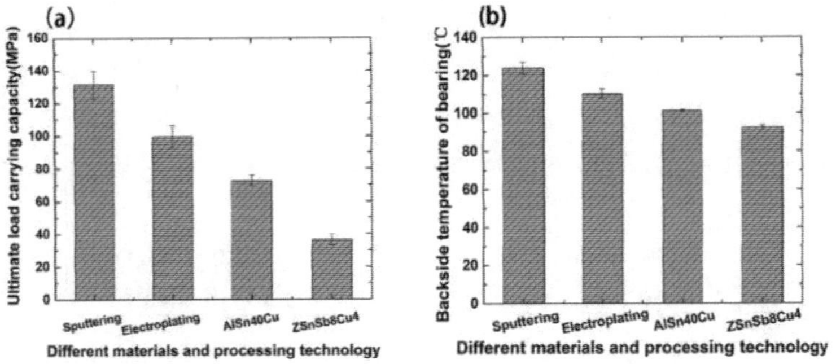

Figure 8. Average fatigue load and backside temperature of bearing

Fig. 8 shows the average fatigue load and the corresponding backside temperature of bearing with different materials and processing technology. The results showed that the average fatigue load (average value of 3 tests) of CBAS-Bearing, CBPE-Bearing, HTAB-Bearing and TB-Bearing is 131.3 MPa， 99.7 MPa， 72.7 MPa and 36.3 MPa, respectively, and the corresponding backside temperature of bearing is 123.8 °C, 110.2 °C, 101.5 °C and 92.3 °C, respectively. It can be seen from the figure that with the increase of the test load the temperature of the bearing is increased. When the experimental load increased to 27 MPa the backside temperature of TB-Bearing is over 90 °C, which is close to the maximum working temperature of white alloy. Therefore, the bearing capacity will be adversely affected.

4. Conclusion

The ultimate load carrying capacity of journal bearings with different matrix materials and surface treating methods are studied using a bearing fatigue strength test machine in the present work. The following conclusions can be drawn: (1) The copper-based sputtering bearing has the best bearing capacity, and the ultimate bearing capacity reaches 131.3 MPa. (2) Surface treatment has a significant effect on the wear resistance of the bearing, the wear resistance of sputtering coating is better than that of electroplating coating. (3)The main failure form of CBAS-Bearing and CBPE-Bearing is abrasion wear of sputtered or electroplated coating, while the failure form of HTAB-Bearing and TB-Bearing is fatigue of alloy layer.

Acknowledgements

The authors acknowledge the financial support provided by Marine Low Speed Engine Project-Phase I (Grant No. CDGC01-KT11).

References

[1] Zhang Y, Tudela I, Pal M, Kerr I. High strength tin-based overlay for medium and high speed diesel engine bearing tribological applications. Tribology International. 2016;93:687-95.
[2] Ji X, Chen Y. Tribological Behavior of Babbitt Alloy Rubbing Against Si3N4 and Steel Under Dry Friction Condition. Journal of Materials Engineering and Performance. 2016;25:750-5.
[3] Cristea A-F, Bouyer J, Fillon M, Pascovici MD. Transient Pressure and Temperature Field Measurements in a Lightly Loaded Circumferential Groove Journal Bearing from Startup to Steady-State Thermal Stabilization. Tribology Transactions. 2016;60:988-1010.
[4] Gebretsadik DW, Hardell J, Prakash B. Friction and wear characteristics of different Pb-free bearing materials in mixed and boundary lubrication regimes. Wear. 2015;340-341:63-72.
[5] Bican O, Savaşkan T. A comparative study of lubricated friction and wear behaviour of Al–25Zn–3Cu–3Si bearing alloy. Proceedings of the Institution of Mechanical Engineers, Part J: Journal of Engineering Tribology. 2014;228:896-903.
[6] Senthil Kumar P, Manisekar K, Subramanian E, Narayanasamy R. Dry sliding friction and wear characteristics of Cu-Sn alloy containing molybdenum disulfide. Tribology Transactions. 2013;56:857-66.
[7] Miguel J, Guilemany J, Dosta S. Effect of the spraying process on the microstructure and tribological properties of bronze–alumina composite coatings. Surface and Coatings Technology. 2010;205:2184-90.

IOP Conf. Series: Materials Science and Engineering **689** (2019) 012015 doi:10.1088/1757-899X/689/1/012015

Electromechanical Device for Temperature Control of Internal Combustion Engines

A Mamani[1], G Quispe[2], C Raymundo-Ibañez[3,*]

[1] Facultad de Ingeniería, Universidad Continental, Lima 15046, Perú.
[2] Ingeniería Industrial, Universidad Nacional Autónoma Altoandina de Tarma, Tarma 12651, Perú.
[3] Dirección de Investigación, Universidad Peruana de Ciencias Aplicadas (UPC), Lima 15023, Perú.

*carlos.raymundo@upc.edu.pe

Abstract. Internal combustion engines are the most commonly used engines in the automotive world. However, these engines lack an overheating prevention system against cooling system failures when they exceed their normal operating temperature. Less experienced drivers (users) usually do not notice overheating until the engine stops, generating economic expenses in engine repairs. As such, this paper describes the design and construction of an electromechanical device to prevent engine overheating. This device is installed in a vehicle and operates independently from the electronic control unit (ECU); it records the coolant temperature and controls air admission to the engine of the vehicle in which it is installed. In addition, a new Arduino-based card will receive signals from a temperature sensor as input and process them according to its programming. Then, it will send signal outputs to the actuators: a servomotor, monitor, LED display, and buzzer. To control the intake flow, a butterfly valve is used with the servomotor. This valve partially or totally restricts the engine airflow, based on the temperature programmed for the Arduino, thus protecting the engine from overheating.

1. Introduction

The overheating of internal combustion engines causes lubrication problems due to lowering oil viscosity levels. Further, some parts can be deformed. Some factors that might cause the generation of high temperatures in a water-cooled internal combustion engine are as follows: obstruction of the air circulation holes of the radiator, water pump failures, radiator-fan-belt breakages, radiator water leaks, stuck thermostat, and other cooling system related issues [1]. Because engines may reach impressive internal temperatures of up to 2000 °C, the cooling system is designed to disperse part of the temperature produced by the combustion process of the engine. Therefore, it must absorb, circulate, control, and dissipate temperature [2]. Prompted by competitive market needs, the modern automotive industry faces the problem of having to develop and implement in its products the changes produced by technological evolution in amounts of time that are becoming increasingly smaller and with greater budget constraints. In fact, only modeling and virtual simulation can help engineers to respond with the required swiftness and at the lowest costs possible. Apart from boosting responsiveness, modeling and simulation tools offer the advantage of streamlining changes and evaluations of the proposed new designs [3]. Water cooling means that vehicle cylinders are housed inside a casing that is filled with water circulated through a pump. Then, water is cooled when it passes through the radiator blades. It is important for the engine-cooling liquid to not be ordinary water because combustion engines work regularly at temperatures exceeding the boiling temperature for water [4]. Additionally, feedback control systems are called closed-loop control systems. In practice, the terms feedback control and closed-loop control are used interchangeably. In a closed-loop control system, the actuation error signal is fed to the controller. This error is the difference between the input and feedback signals (which can be the output signal or based on a function of the signal output and its derivatives and/or integral functions). The error is fed into the controller to reduce the error and force the system output to return to an expected value [5]. Temperature sensors are also called thermistors. The resistance of conventional metal-oxide thermistors decreases non-linearly, which means that when temperature increases, their resistance decreases. Such thermistors have negative temperature coefficients (NTC), although there are also thermistors with positive temperature coefficients. The resistance change for each degree of temperature alteration is much greater than the change that occurs with metals [6]. When dosing air to the engine, cylinders are filled by aspiration, during the piston downstroke while the intake valve is open, or by previously compressing air, which is called overfeeding. Unlike a diesel engine, the Otto engine can only regulate loads by the amount of mixture introduced and burned in the cylinder. Since the dosing must be continuous, the per-cycle work depends directly on the air mass. In other words, the power developed is a function of mass airflow.

Airflow is regulated using a butterfly valve placed in the intake manifold and operated by the driver when the accelerator pedal is pushed. The butterfly valve creates a variable pressure drop so that upstream, the existing pressure is approximately equal to the atmospheric pressure, while downstream, it will always be lower [7]. Invented by James Watt in the eighteenth century, the butterfly valve is one of the most popularly used valves that currently exist as its design facilitates its installation, maintenance, and operation, thus generating savings and benefits for a company. It has a body, which comprises a disk that rotates 90° around an axis and controls the flow of liquids or gases [8].

Arduino is an open-source electronic prototype platform based on flexible and easy-to-use hardware and software. It was created for students, designers, and anyone interested in developing interactive objects or environments. Arduino can interpret the environment by receiving its inputs from a wide variety of sensors and can modify its environment by controlling devices such as transducers, lights, and motors. The board's microcontroller is programmed through the "Arduino Programming Language" (based on Wiring1) and the "Arduino Development Environment" (based on Processing2). Arduino projects can be stand-alone or they may interface with software running on a computer, tablet, or mobile device. The plates can be assembled by hand or ordered preassembled. The software can be downloaded for free. The reference hardware designs are available based on an open-source license so they can be adapted to different needs [9].

2. Methodology and design

2.1. Proposed design

The device is designed with two components: an electronic component and a mechanical component.

2.1.1. The electronic component records, processes, and provides immediate signal responses to actuators. Before assembling the circuit, it was necessary to use the Proteus software to create a simulation that would determine whether the electronic system was working properly. Then, the electronic layout will be created to assemble the electronic system, as shown in Figure 1. The proposed electronic card is based on an Arduino module. The voltage supply is 5-V direct current, which will be regulated using a voltage regulator. Further, the NTC temperature sensor will emit signals as input for the Arduino module, which is in charge of processing the information and of providing responses as output signals to actuators. The responses can include that the servo motor will move at a certain angle and activate the relay to press the buzzer at the maximum temperature programmed. An LCD monitor (screen) will display the current coolant temperature and position of the servomotor.

With all these data, the design of referred components for the assembly is executed. With the layout, a card can be burned with the circuit designed to place the components based on the Arduino module used. Then, the components will be welded using tin and a soldering iron to prevent them from moving.

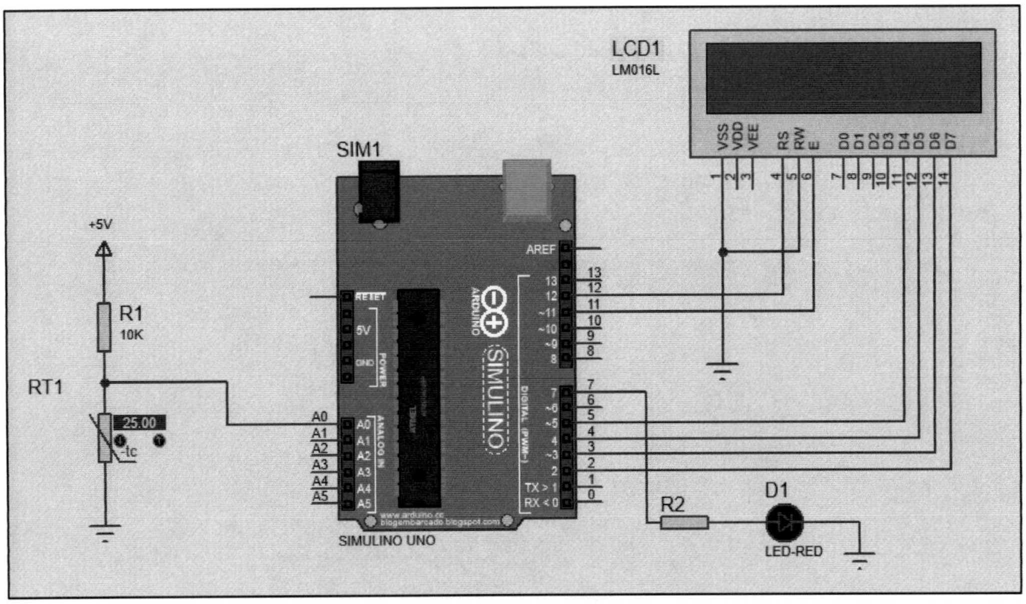

Figure 1. Simulated circuit in Proteus

2.1.2. The mechanical component is included in the design of the butterfly valve to restrict airflow, for which it is necessary to calculate cubic feet per minute (CFM) operating motor parameters (cubic inches per min) and the valve torque.

CFM calculation
Data: for a 1300-cc 2nz engine
Compression Ratio = 10.5 to 1
Maximum power = 63 KW (84.5 HP) 6000 rpm
Maximum torque = 121 Nm (12.3 kgf.m) 4400 rpm

Table 1. Calculation of airflow for internal combustion engines

Number of cylinders	Cubic centimeters	1,300	
Revolutions per min (RPM)	Enter the highest expected value	6,000	
Volumetric efficiency	Non-turbo gasoline engine	3	0.8
Results	CFM		110
Liters per min of air required for a maximum speed			3,120

Note: the calculation is based on a maximum consumption for a turbocharged gasoline engine. The values of certain electronic controls can increase this air value by up to 2.

With maximum values to increase the design factor

$$CFM = 110\ ft3/min \qquad Q=0.0519142195 m3/seg \tag{1}$$

Speed and flow

$$A = \pi r^2 = \pi\ 0.0295^2\ m2 = 2.64\ m2 \tag{2}$$

Flow

$$Q = VA$$

Calculation of V (speed)

$$V = Q / A = 0.05/2.64 = 0.018\ m/s \tag{3}$$

Valve simulation

The butterfly valve will be installed at the intake inlet between the filter and vehicle's throttle body. In Figure 2, the airflow can be noticed when the valve is completely open, and the atoms are heated when passing through the valve.

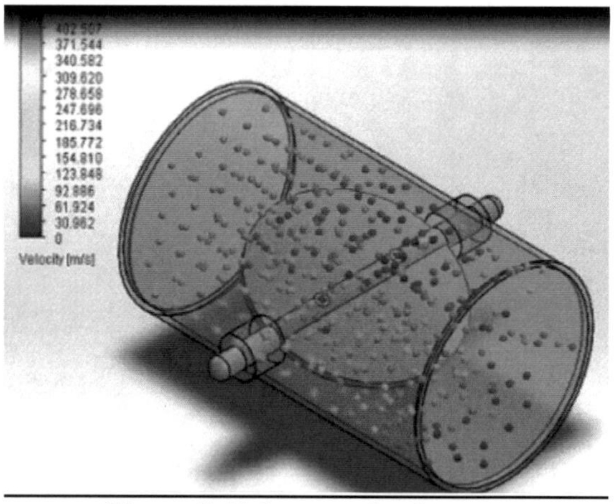

Figure 2. Valve completely open

This principle will be used to restrict engine airflow to control overheating. The servomotor will be installed on the valve shaft and controlled by the Arduino module because the temperature range of an engine, for normal operations, is 82 °C–96 °C. At 82 °C, the thermostat opens; if the temperature is lower, it closes. In this way, the temperature is maintained above 82 °C. When the temperature rises to 96 °C, the fan is activated and cools the radiator. Thus, the engine maintains its working temperature range [10]. To ensure that temperature is not exceeded (for a 0 °C–96-°C range), in its programming, it is requested for the servomotor not to move. In this case, the valve is left in a horizontal position and completely open. When the engine reaches 97 °C, the device works by moving the valve to the 45°-counterclockwise position, restricting the airflow and partially reducing power and torque. If the driver was stepping on the accelerator to achieve the maximum speed and the engine was operating at a high number of revolutions, the normal operating temperature range would have been exceeded. It will not be possible to continue stepping on the accelerator owing to the lack of air, and when the engine temperature is lowered to its working range, it will return to normal. If the temperature increases (and reaches 98 °C–99 °C), the valve will move an additional 45°, placing itself at 90°. In this position, it will be completely closed, as shown in Figure 3. The engine will stop and be turned off owing to a lack of air. In this way, the device will fulfill its objective of stopping the mechanism and preventing the engine from overheating.

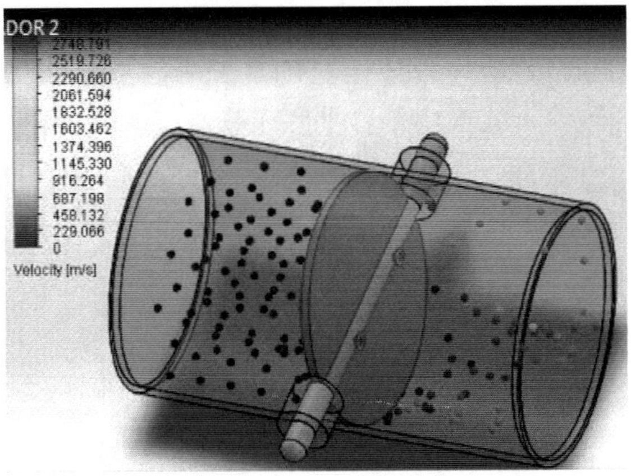

Figure 3. Valve close

When the engine cools and temperature drops to below 96° C, the engine will return to normal; the valve will open completely and the device will once again automatically work with the other components, thus fulfilling a closed-loop control with feedback. If the device was not installed and the water pump was not working, the engine would overheat at high temperatures. This is when the valve would perform the main functions for which it was designed: a perfect closure and ensuring that the axis is not deformed or broken. This is the reason why a static stress analysis was performed, as shown in Figure 4.

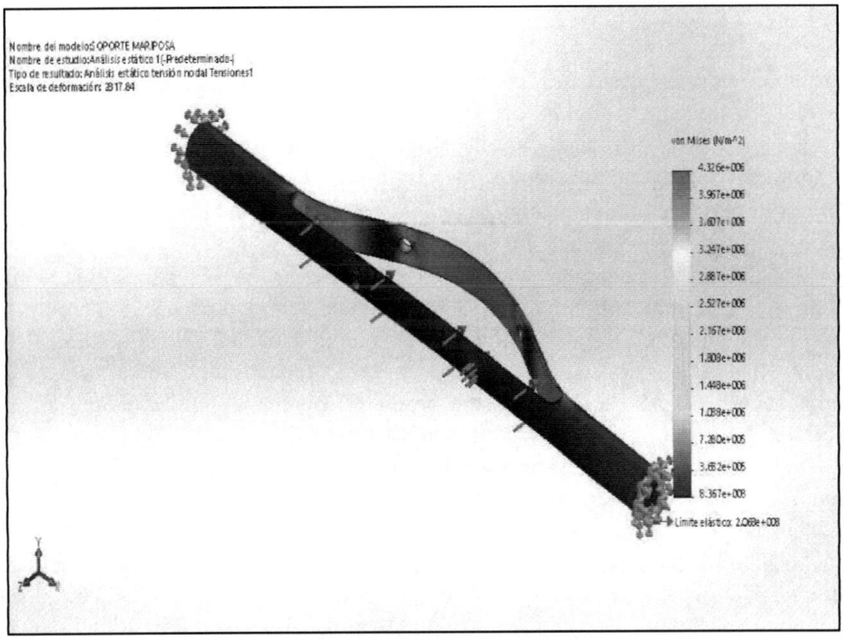

Figure 4. Static stress analysis

3. Results and analysis

An automotive scanner was required to read the temperature sensor, and a multimeter was used to experimentally obtain field data.

The temperature sensor is a negative-temperature-coefficient (NTC) sensor. When the temperature rises, resistance decreases.

IOP Conf. Series: Materials Science and Engineering **689** (2019) 012015 doi:10.1088/1757-899X/689/1/012015

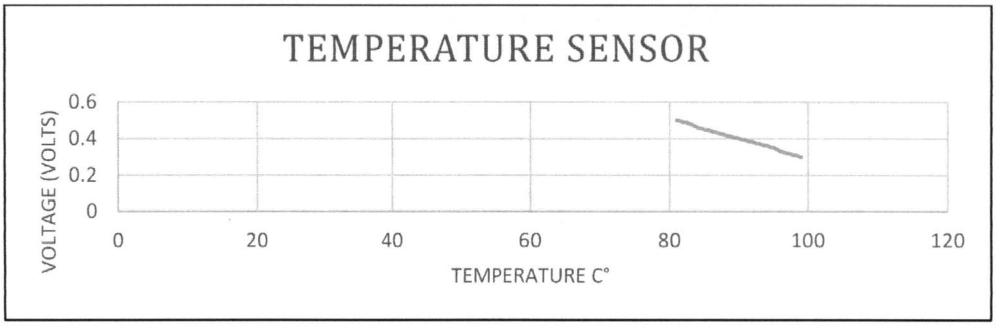

Figure 5. Temperature Sensor

The static analysis simulation software results for the butterfly valve shaft are as follows:

Table 2. Static analysis simulation software

Selections set	Units	Sum X	Sum Y	Sum Z	Results
The complete model	N	7.45058e-008	-3.85568e-007	0.999987	0.999987

As shown in Figure 5 and in the results from Table 2, the lowest resulting force occurs along the Z axis. Further, the simulation depicted in Figure 4 does not exhibit any red surface, which means that the axis has no stress and diameter and selected material are correct. The butterfly valve can be manufactured according to the designs produced by Solidworks and design parameters.

4. Conclusions

- Electronic components and their programming are the easiest parts, but the design of the mechanical component is slightly complex.
- Regarding the valve, programs such as Solidworks are quite necessary and useful tools for simulation and to draw up the layouts.
- When developing the device prototype, a set of processes and steps are listed for the project, which we have to follow as they are structured in a sequence.
- When designing the device, many factors, such as the engine shutdown, were taken into account. At first, this only involved a thermoswitch and power cutoff relay in the ECU. However, after careful analysis, other problems might occur. For example, if the vehicle is traveling at a high speed and turned off at any moment, the brakes would not function properly owing to the vacuum booster produced by the engine that causes the brakes to become rigid. Steering, on many models, is hydraulically powered and if the engine does not work, the hydraulic steering system would not work either. These problems could cause very serious accidents. However, in terms of safety, the proposed design would not bring about such problems.
- When assembling the electrical components, we could not find a temperature sensor for the Arduino module in the market. Therefore, the module had to be tested using a sensor similar the one used for the engine. This sensor worked well after a voltage divider was installed as a modification.

References

[1] Fernández-Morales, Flavio Humberto and Duarte, Julio Enrique. Módulo para la medición de variables en motores de combustión interna. Ingeniería y Tecnología. 2013. Vol. 9, no. 2, p. 250–256.
[2] AUTOMOTRIZ, Escuela D E Ingeniería. Chimborazo (2011).
[3] Romero, Carlos, Carranza, Yamid and Quintero, Héctor. Modelado del calentamiento de los motores de combustión. . 2007. No. 37, p. 249–254.

[4] Cordova Morales, Edison. Estudio de un motor de combustión interna para determinar sus parametros de funcionamiento y su factibilidad de aplicacion en los laboratorios de la facultad de Ingenieria civil y mecánica. 2012.

[5] Ogata, Katsuhiko. Ingeniería de Control Moderna. 2013. ISBN 9788578110796. Applicability for this approach.

[6] Bolton, William. Mecatrónica, Sistema de control electrico en la ingeniería mecánica y eléctrica. [no date]. ISBN 9786077076032.

[7] FUNDAMENTOS DEL CONTROL DE MEZCLA. [online]. Available from: https://www.tdx.cat/bitstream/handle/10803/5933/2Fundamentos_control_mezcla.pdf?Sequence=3&isallowed=y

[8] Luis Felipe Villarroel Quinde (PONTIFICIA UNIVERSIDAD CATOLICA DEL PERU). Diseño de válvula de admisión tipo mariposa con diámetro nominal de 750 mm para una central hidroeléctrica de 34.7 m de salto neto (2013).

[9] Perea Palacios, Jhonny Wbeimar. Diseño De Un Sistema De Monitoreo, Registro Y Control De Temperatura Y Humedad Para Un Cultivo De Invernadero. Universidad Tecnológica De Pereira [online]. 2016. P. 79. DOI 10.1056/NEJM199405193302003. Available from: http://repositorio.utp.edu.co/dspace/bitstream/handle/11059/7436/621317P434.pdf?Sequence=1 Después de controlar el microclima del invernadero, se almacenan en una base de datos las variables y detalles más importantes para garantizar un historial de las variables de estado que el agricultor desea analizar. Con esta base de datos, se tiene una herramienta para optimizar las cosechas y garantizar control predictivo

[10] TOYOTA MOTOR CORPORATIONR. 1NZ-FE ENGINE.

IOP Conf. Series: Materials Science and Engineering **689** (2019) 012016 doi:10.1088/1757-899X/689/1/012016

Heavy Object Lifting Platform to Correct Human Balance and Posture

F Raymundo[1], G Quispe[2], C Raymundo-Ibañez[3,*]

[1] Facultad de Ingeniería, Universidad Continental, Lima 15046, Perú.
[2] Ingeniería Industrial, Universidad Nacional Autónoma Altoandina de Tarma, Tarma 12651, Perú.
[3] Dirección de Investigación, Universidad Peruana de Ciencias Aplicadas (UPC), Lima 15023, Perú.

*carlos.raymundo@upc.edu.pe

Abstract. This research document on technological development is aimed at workers in small companies, those that are developing well and soon to emerge as companies that will be positioned at the top of their labor sector. Starting from the problem at present with respect to the bad manipulation of loads it is arranged to solve the problem with established objectives to try to correct the corporal posture at the time of lifting an object with a machine that replaces the functions of the worker. The design and the solution was made based on the requirements of the worker, workers who have the need to load objects at a certain height, here the man-machine relationship, demands and desires that help to achieve the general objective are analyzed. In addition, it has the main function of this machine: to lift heavy objects of up to 150 kg of mass, this due to the implementation of a hydraulic piston in the lifting platform. The appropriate solution is an adaptation of a platform in the form of a truck forming an angle of 47.17 between the structure of the truck and the horizontal surface, this position is the most adequate so that the back does not suffer injuries during the handling of loads at work. The results give assurance that it is a machine of reliability and easy handling since it only requires an operator to control the objects and a flat position on the back without damaging the spine.

1. Introduction

In certain work areas, lifting heavy objects require human effort, which may cause excessive damage to the lumbar area owing to the lack of knowledge and application of new technologies, particularly in small businesses (microenterprises). Therefore, this study aims to apply knowledge, skills, and techniques to address this issue and help workers at small booming companies, which may soon emerge as companies that will position themselves at the top of their industrial sector. Using the existing poor load handling issues as a point of departure for problem resolution, the objectives are set to correct body posture using a machine to perform all the heavy lifting tasks for the workers.

Therefore, this study uses the VDI 2221 methodology, which describes all the tools used to obtain one or multiple solutions, considers the list of requirements, explains their operation, and calculates the effort needed to lift objects and improve human work efficiencies. In the future work, this study will be responsible for implementing the technology and automation required for new jobs based on automatic control of lifting platforms, thereby replacing human effort with mechanical effort and preventing posture-related injuries resulting from lifting heavy objects.

2. State of the Art

2.1. Human posture analysis

2.1.1. Symmetry

To focus on poor posture acquired by workers when asymmetrically lifting heavy jobs at the work areas, previous studies addressed the lumbar spine compression and shear forces because of poor posture and determined that these forces majorly affect the lumbar spine during asymmetric lifting. [1] In storage, transportation, and logistics sites, where companies require workers to move, load, and lift heavy objects regularly without adequate rest, the asymmetric lifting or handling of loads helps them to reduce back pain. Assessing the relation between manual heavy object lifting and the stress exerted by compression and shear forces on the lumbar area revealed that the difference between a rookie and an experienced worker relies on knowledge, i.e., experienced workers have better knowledge; therefore, they assume better postures when loading objects. [2] (Fig.1).

The results have been measured and compared against each other, which indicated the symmetry model and angle when lifting heavy objects. Regarding the difference between the lifting techniques used by experienced workers and rookies, the rookies apply greater effort than experienced workers, which is a global issue. [3] Eventually, workers start understanding load differences, thereby performing healthy work particularly without accidents or lumbar injuries resulting from lifting heavy objects. [4] (Fig. 2).

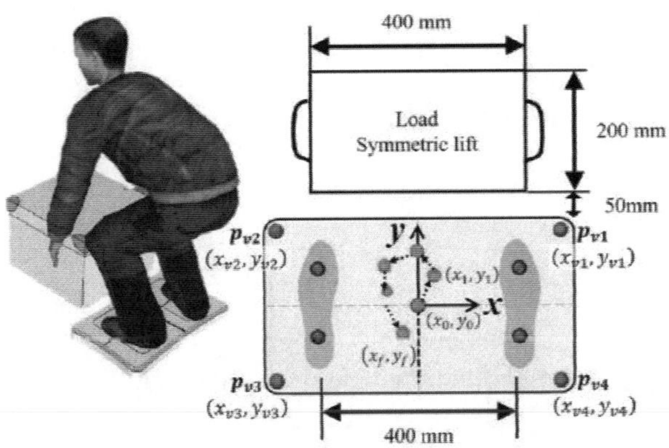

Figure 1. Experimental Symmetry in Environments used in COP Measures

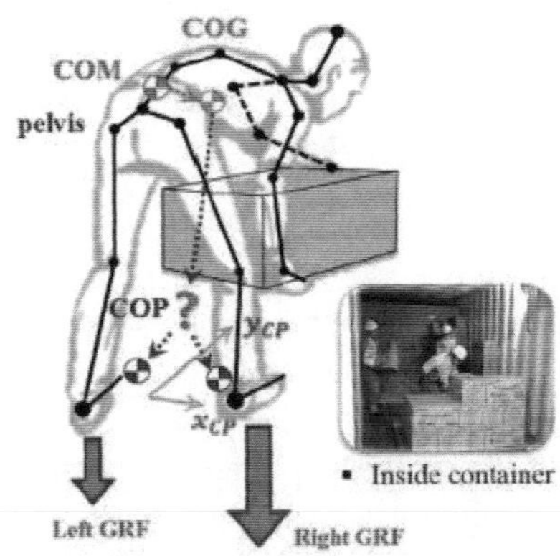

Figure 2. Symmetry Analysis through COP Locations

2.1.2. Balance

Technological innovations have introduced robot prototypes for people with serious back issues resulting from poor lifting practices for heavy objects. Evidently, treatments had not been effective, although medicine and technology made several efforts to help the people who want to get up from bed through a metal robot support, which may understand and care for patients. This is a controversial issue because some people assert that robots must not take care of patients [5].

Back pain is the most frequent affliction in the world, often requiring a doctor or hospital care. Several doctors, employers, and employees have started using lumbar support belts to relieve some of these issues. However, the problem regarding the cost remains because numerous support belts are not very effective in eliminating or reducing back pain, particularly because employers often purchase the cheapest belt available owing to the lack of technical knowledge, which exposes them to being ripped off.

Further, assessments on the effects that are exerted on the waist by carrying heavy objects revealed that the balance is critical when moving objects, together with using adequate lumbar support belts while working. Moreover, the most affected workers, who are more likely to suffer back injuries due to poor balance and posture when loading heavy objects, are Wal-Mart employees. [6] (Fig. 3)

3. Contribution and proposed solution

This study proposes combining three machines in an unusual design to improve ergonomics and operator posture control. [7] Herein, hydraulic pistons are widely used in big cranes at work areas of large magnitude, and platforms serve to support and lift loads. Finally, the wheelbarrows currently used by microenterprises usually cause bad posture because they are used to move large and heavy objects, which eventually turn into additional weight. (Fig. 4)

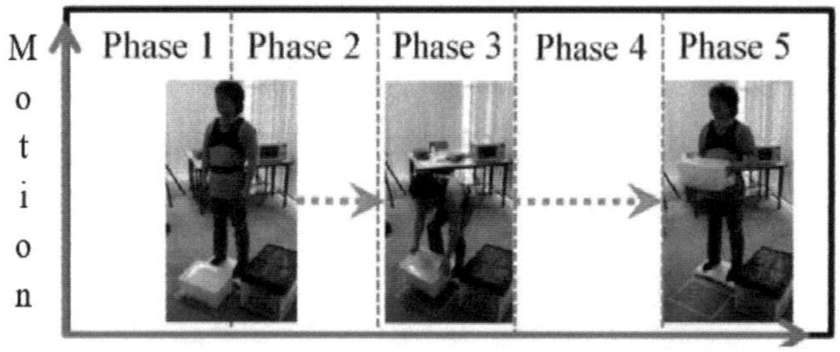

Figure 3. Experimental Study Process

Figure 4. Load Lifting Platform Design.

The alternative solution proposed herein is the design of a new lifting platform, which comprises three main components: wheels, metal structure, and a hydraulic piston. This new platform is collapsible, can reach up to 2 m in height, and provides stability to lift any type of heavy objects. Its design weight is 130 kg and comprises the aforementioned three components currently used. Additionally, the platform does not require much space and is the most suitable for working in small spaces. However, it requires an operator because it will be managed by a semi-automatic control guided by the operator, thereby replacing human force with hydraulic force.

3.1. Operation

From Figure 5:

- Preparation: Object identification.
- Execution: Moving the object from point A to point B, as required by the programmer, such as operator and worker.
- Control: The procedure can be controlled through the buttons and command controls of the lifting machine.
- Final Phase: The work ends when the object is delivered to the target location.

3.2. Operation sequence

From Figure 6:

- Start: The machine uses electronic, electric, and manual controls to lift and move objects.
- Object identification: The operator identifies the object that requires lifting and moving from a specific location to another location and identifies the type of platform (i.e., basket) required.
- Equipment positioning: The operator positions the machine using the command controls and then puts the platform in place to be used for the load.
- Object loading: The operator secures the load on the platform and then proceeds to operate the machine again.
- Transportation: Moves the object from the original location to another specific location.
- Storage: Drops the load safely in the desired location and starts returning the machine.

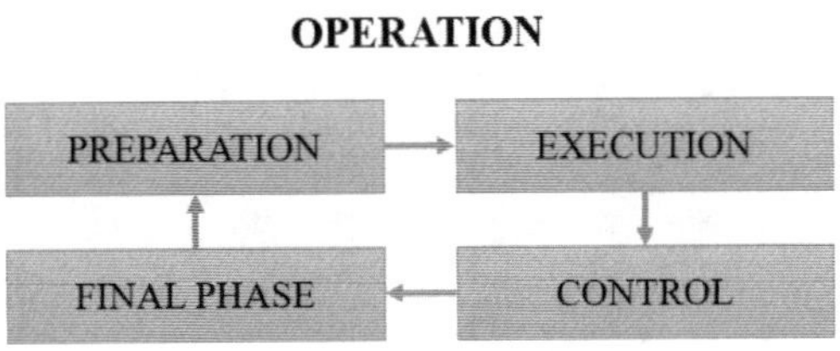

Figure 5. Operation

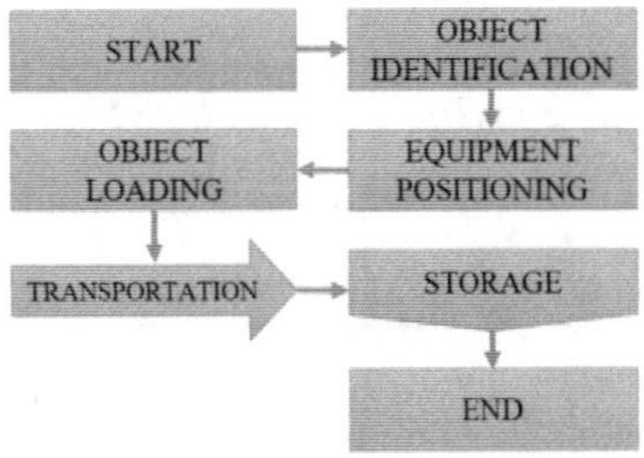

Figure 6. Operation sequence

3.3. Calculations

3.3.1. Determination of the angle between the structure and horizontal surface

From Figure 7, herein, Structure Measurement = 1.50 m; Average people height = h = 1.10 m. Then:

$$Sin\beta \;= h/(Structure\ Measurement) \;=\; 1.10/1.50\,;\; \beta \;=\; 47.17\;^{\circ} \tag{1}$$

3.3.2. Effort at maximum load

The calculations consider the maximum load to obtain favorable results (Fig. 8).

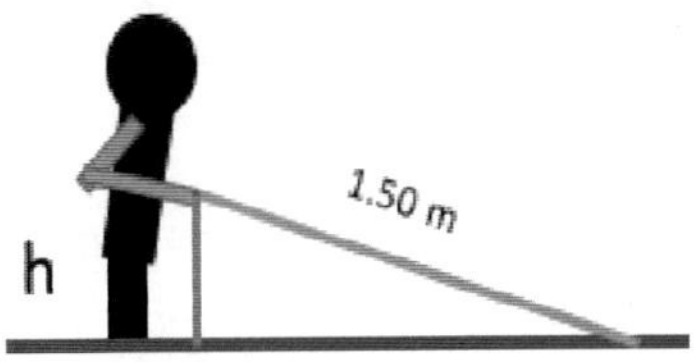

Figure 7. Angle between the structure and horizontal surface

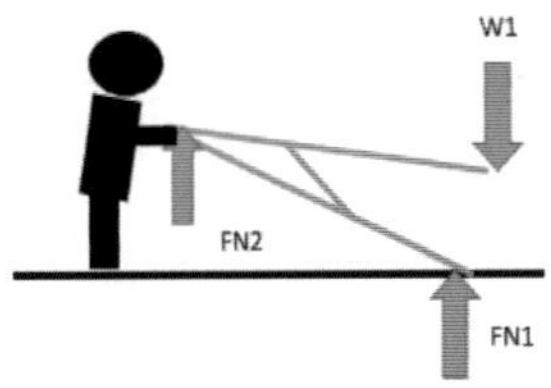

Figure 8. Effort at maximum load

Second Law of Newton: At equilibrium, the sum of the forces acting on the object equals zero.

Σ F (Y)=0; where W1 = 80 kg.; Then:

$$-Fw1 - Fw2 \;+\; Fn1 \;+ Fn2 \;=\; 0;\; Fn1 \;+ Fn2 \;=\; Fw1 \;+\; Fw2 \tag{2}$$

$$F = m * g \tag{3}$$

Replacing data:

$$Fw1 = 80\;kg * 9.8182\;m/s\text{^}2 = 784{,}8\;N;\quad Fw2 = 50\;kg * 9.8182\;m/s\text{^}2 = 490.91\;N \tag{4}$$

Then, replacing in (2)

$$Fn1 \;+ Fn2 \;=\; Fw1 \;+\; Fw2 \;=\; 784{,}8 \;+\; 490.91 \;=\; 1275.71\;N$$

To calculate the force exerted on the piston to reach equilibrium (Fig. 9),

$$\Sigma F(Y) = 0; \qquad -Fw1 - Fw2 + Fn1 + Fn2 + Fpsen30 = 0;$$
$$Fn1 + Fn2 + Fpsen30 \;= +Fw1 + Fw2 \tag{5}$$

We will find the starting force to determine the force exerted on the piston:

$$\Sigma Fpw(Y) = 0\,;\quad Fpw \;= 7348;$$
$$\Sigma Fpipw(Y) = 0 \quad;\quad -Fpwsen30 - Fpsen45 \;=\; 0;$$
$$Fp\;sen45 \;=\; Fpwsen30;\quad Fp \;=\; 5195\;N;$$

Then, it is replaced in

$$Fn1 \;+ Fn2 \;+ Fpsen30 \;=\; +Fw1 \;+\; Fw2; \tag{6}$$
$$Fn1 \;+ Fn2 \;=\; -3920N$$

System of equations between (4) and (6)

$$Fn1 \;+\; Fn2 \;=\; 1275.71; -Fn1 \;+ Fn2 \;=\; -3920N$$
$$F1 \;=\; -1322;\; F2 = \;2597 \tag{7}$$

3.3.3. Piston speed in the cylinder

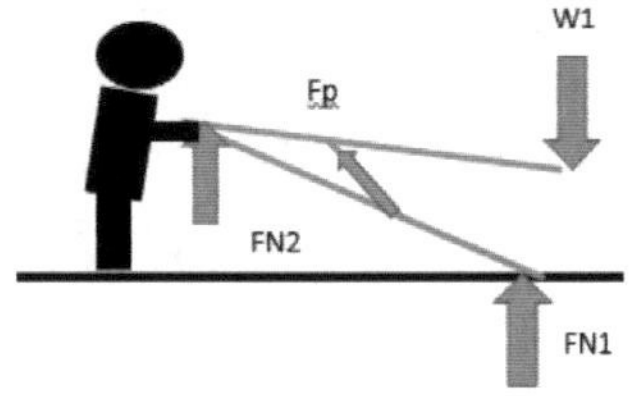

Figure 9. Force exerted on the piston to reach equilibrium

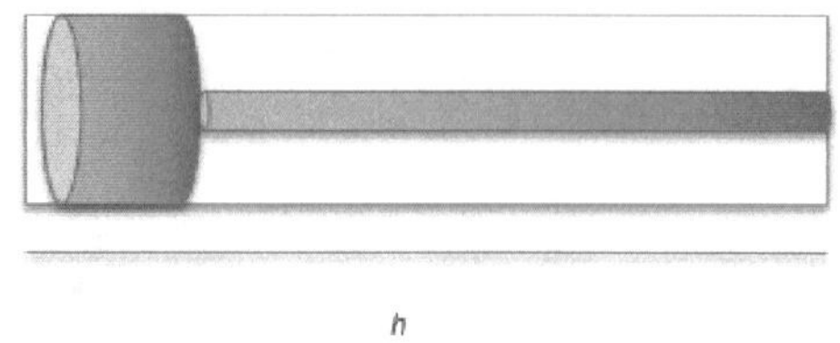

Figure 10. Piston speed in the cylinder

$$P = F/A \tag{8}$$

Development of pressure in the cylinder: F = 2550N; A=2pulg <> 0,0508

$$F = P * A \quad ; \quad 2550 = P * (\pi * d\text{^}2)/4 \quad ; \quad P = 12\,4827{,}05 \approx 20PSI \tag{9}$$

NOTE. This is the piston pressure at equilibrium. For upwards movements, an additional 25% pressure must be supplied so that the piston may move the load. We will use the simple rule of three:

$$X = (2550N * 25)/100 = 637{,}5\ N \tag{10}$$

Total force required to lift 80 kg in weight: F = 3187.5 N at P = 40 PSI
Finally, the speed of a cylinder

$$V = v/A \tag{11}$$

Where: V = Speed (m/min); v = Volume (m^3); A = Area (m^2)

$$V = \frac{\pi * r^2 * h}{\frac{\pi * d^2}{4}} \tag{12}$$

h = 50 cm <> 0.5 m ; dp = 2pulg <> 0.0508

$$V = \frac{\pi * \left(\frac{0.051}{2}\right)^2 * 0.5}{\frac{\pi * 0.051^2}{4}} = 0.5\frac{m}{min} <> 50\ cm/min$$

The calculated speed is as expected because as engagement time decreases for the workers, productivity increases.

4. Simulation

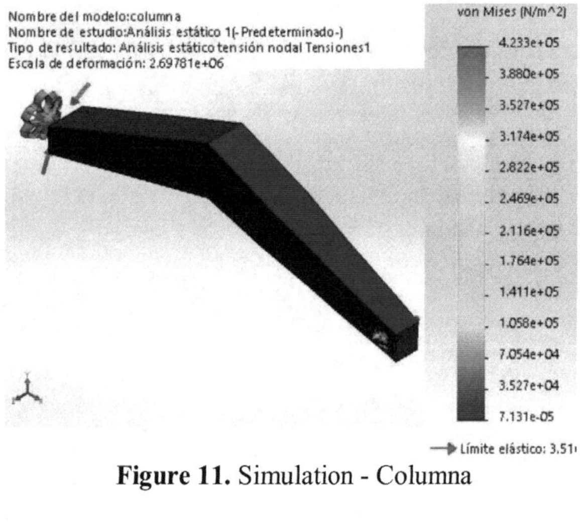

Figure 11. Simulation - Columna

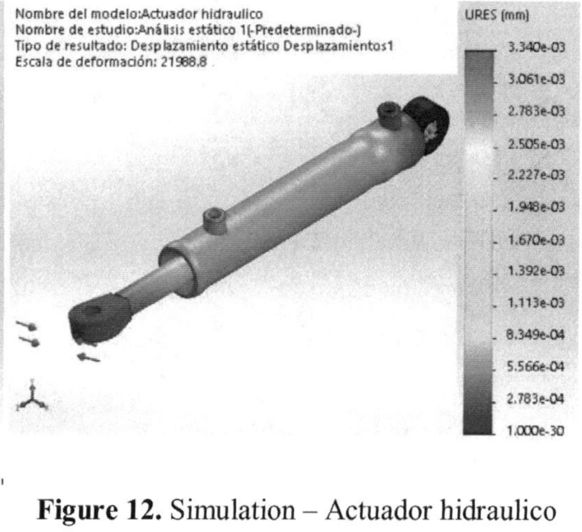

Figure 12. Simulation – Actuador hidraulico

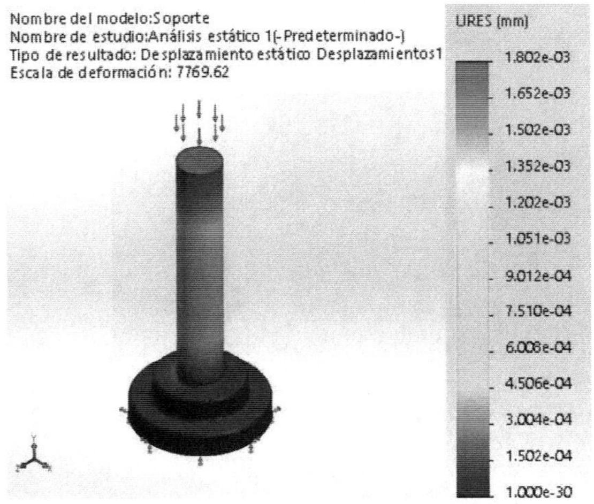

Figure 13. Simulation - Soporte

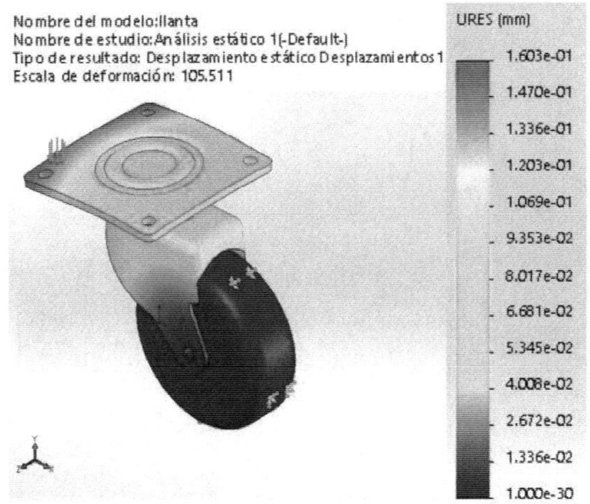

Figure 14. Simulation - Llanta

5. Conclusions

- The best lifting mechanisms are operated at high pressure, whether hydraulic or pneumatic, using extremely high compressed pressure to lift heavy objects to a certain height, thereby reducing engagement time for workers.
- Warehouse dimensions are often restricted in microenterprises, which prompted the design of a machine with specific dimensions that can replace one of the workers.
- The material used to store heavy objects is wood because its Britney hardness is 1 to 7 HB. However, this material is weak during several environmental conditions, such as rain. Additionally, it bumps when moving loads, thereby causing it to lose its original properties.

References

[1] JEONG, H. y OHNO, Y., 2017. Symmetric lifting posture recognition of skilled experts with linear discriminant analysis by center-of-pressure velocity. Intelligent Service Robotics, vol. 10, no. 4, pp. 323-332. ISSN 18612784. DOI 10.1007/s11370-017-0227-8.

[2] JEONG, H., YAMADA, K., KIDO, M., OHNO, Y., WATANABE, S. y NOMURA, T., 2016. Analysis of Center of Pressure Location during Asymmetric Lifting. Proceedings - 2015 IEEE International Conference on Systems, Man, and Cybernetics, SMC 2015, pp. 2471-2474. DOI 10.1109/SMC.2015.432.

[3] WATANABE, S., TORIGAI, K., HAYASHI, M., JEONG, H., SHIMIZU, S., KIDO, M., NOMURA, T., MIYOSHI, E., YAMADA, K. y OHNO, Y., 2015. A difference of human posture between beginner and expert during lifting a heavy load. BMEiCON 2014 - 7th Biomedical Engineering International Conference, pp. 1-5. DOI 10.1109/BMEiCON.2014.7017388.

[4] JEONG, H., YAMADA, K., KIDO, M., OKADA, S., NOMURA, T. y OHNO, Y., 2016. Analysis of Difference in Center-of-Pressure Positions between Experts and Novices during Asymmetric Lifting. IEEE Journal of Translational Engineering in Health and Medicine, vol. 4, no. August 2015. ISSN 21682372. DOI 10.1109/JTEHM.2016.2599185.

[5] WANG, T., JEONG, H. y OHNO, Y., 2017. Evaluation of Self-Reliance Support Robot Through Relative Phase. IEEE Access, vol. 5, pp. 17816-17823. ISSN 21693536. DOI 10.1109/ACCESS.2017.2747841.

[6] SHIMIZU, S., JEONG, H., KIDO, M., MIYOSHI, E., YAMADA, K., OHNO, Y., NOMURA, T., WATANABE, S., TORIGAI, K. y HAYASHI, M., 2015. Effectiveness evaluation of waist support tool through human posture balance. BMEiCON 2014 - 7th Biomedical Engineering International Conference, DOI 10.1109/BMEiCON.2014.7017387.

[7] VASQUEZ TORRES, E.L., 2013. SAE 1020 y SAE 1045. [en línea], pp. 6. Disponible en: https://repository.unilibre.edu.co/bitstream/handle/10901/7826/VasquezTorresEdwinLibardo2013 Anexos.pdf?sequence=2.

IOP Conf. Series: Materials Science and Engineering **689** (2019) 012017 doi:10.1088/1757-899X/689/1/012017

Ventilation System Simulation Model at a Mine

I Artica[1], G Quispe[2], C Raymundo-Ibañez[3,*]

[1] Facultad de Ingeniería, Universidad Continental, Lima 15046, Perú.
[2] Ingeniería Industrial, Universidad Nacional Autónoma Altoandina de Tarma, Tarma 12651, Perú.
[3] Dirección de Investigación, Universidad Peruana de Ciencias Aplicadas (UPC), Lima 15023, Perú.

*carlos.raymundo@upc.edu.pe

Abstract. This research project modeled and simulated a ventilation system at a Mining Concession, obtaining real-time information regarding the fans used to ventilate the mine. The simulation was developed using the VENTSIM 5 software, which also helped define the number of fans and the operating parameters required, field information, mine dimensions, the mining method, production, geothermal gradient, gas emission, air stream, and air pressure. In addition, the results from software operation revealed the need to open chimneys and use fans. Furthermore, the results also suggest that a specialist dedicated to ventilation and fan maintenance must be hired to formalize mining operations as per the Occupational Safety and Health Mining Regulations from Executive Action No. 024-2016-EM as this Mining Concession is currently operating without license.

1. Introduction

Currently, mine ventilation is important because it guarantees favorable conditions for different operating processes. The consumption of explosives, the number of staff members and the gases found in the rock mass owing to mine depth must be controlled by a ventilation system that may guarantee the necessary air stream and speed to maintain an adequate air flow. In addition, this ventilation system must help control suspended dust, maintain temperature within parameters, and sustain oxygen and gases produced by operations within the maximum permissible limits, thus guaranteeing an optimum development of the mining system and operations in the exploration, development, preparation, and underground exploitation processes [1]. The above is in compliance with the provisions from the Occupational Health and Safety Regulation D.S. 024-2016-EM.

The ELMER 70D Mining Concession is developing its underground mining operations at greater depths, which is causing fresh air distribution issues during their operations, thus hindering worker performance and comfort [2].

The absence of a good ventilation study prevents the execution of large and medium projects; therefore, planning and designing an optimal ventilation system is necessary. The Mining Concession does not have ventilation plans in place, which makes it difficult to assess the real problem. Still, the mine has a main fan that does not provide enough air and constantly breaks down as the management does not know how to select a fan for the different tasks and the corresponding specifications to be met. Similarly, the Mining Concession does not have a preventive maintenance program for the fans. The air stream required by the mine has to be dynamic owing to the continuous work of exploitation and penetration and, as they do not have good planning in terms of ventilation, some negative consequences may arise in terms of production and staff members. Under these circumstances, the planning and design of a ventilation system in the ELMER 70D Mining Concession is unavoidable. Parameters such as the mining method, equipment, production ratio, gas emission, and temperature will be considered for the design and modeling of the ventilation circuit with support from the VentSim 5.0 Software, which determine the air stream and pressure that needs to be distributed throughout the Mining Concession [3].

2. Literature review

Mines are continually expanding in size and depth, leading to increased dependence on localized subsoil resources. The use of underground auxiliary fans is a preferred method to increase and control airflow in work areas [4].

An impact study was conducted on four gold mines at different levels in southern Africa, named A, B, C, and D respectively. The parameters and instrumentation used therein revealed the need for auxiliary fans. In this research, 33 auxiliary fans were used in these mines, which resulted in the implementation of a complete ventilation network system [5].

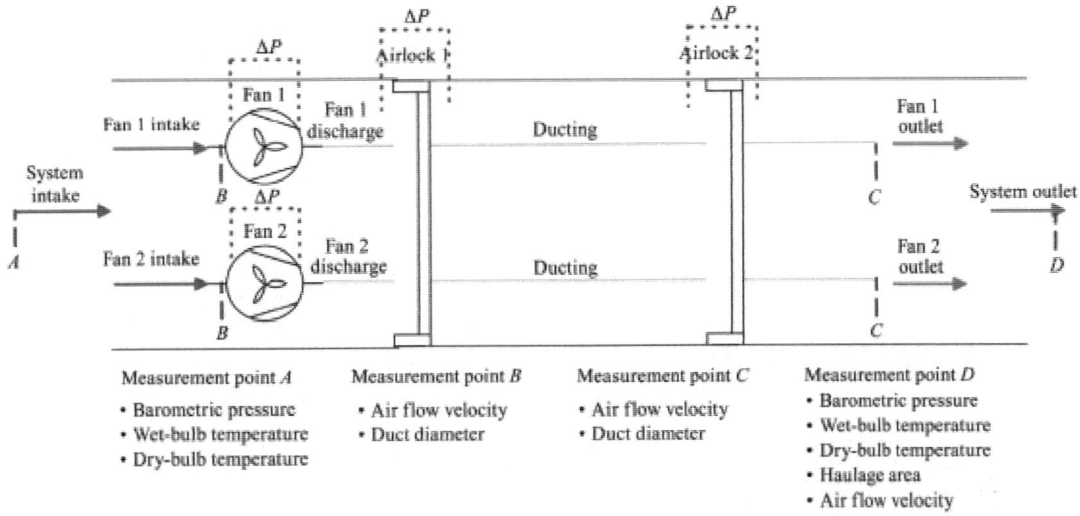

Figure 1. Simulation design of VILLIERES 2019 fans

The ELMER 70D Concession will also deploy auxiliary fans, whose power will be detailed below. In addition, natural ventilation has to be considered, which is the natural moving energy in spite of the geothermal gradient in every underground mine. Therefore, it is ventilation that does not require the use of fans. Mechanical ventilation means that the use of a fan is required to increase airflow; main and secondary ventilation is a ventilation circuit including either Raise Bore, Galleries, or Cruisers. The main fans may be aspirating (when extracting air from the mine) or impelling (when blowing air into the mine). To ensure that all tasks have the amount of air required, redistributing the airflow into the circuit is necessary. One way of achieving this is increasing the resistance of some branch lines through regulators (partitions or doors) or installing secondary fans (boosters). Finally, auxiliary ventilation is used for blind tasks, namely, tasks that have no circuit. The main objective of auxiliary ventilation is to provide, in the front line of work, an airflow enabling the following tasks.

- Adequate dilution of toxic gases.
- Acceptable level of thermal-environmental conditions in the front line of work.
- The air volume required is generally guided through a duct installed inside the task under ventilation. The impelling ventilation is used in our mine, drawing air from the branch of the main network toward the front and then back to the main network from where it was drawn.
- The air shafts in blind tasks must reach a depth of 20 m from the top, thereby providing enough flow to perform the works.

The article by Chang [5] builds a computing model based on physical conditions of a mine front in Western Australia. In the corresponding simulations, the air is considered as non-compressible and

there is no heat transfer based on the study and the gravity, which is 9.8 m/s2. The high-speed air was blown from the exit of the duct and migrated to the direction face in addition to reverse direction airflow after hitting the direction face, and then the apex area between the direction face and the duct head was generated through the combined effect of reverse-direction airflow and the duct air.
For the ELMER 70D Mine Concession, Level 100, the fresh air will not be adequately distributed to shelters, slopes, and sublevels by simply using a compressor and a ventilation shaft because natural ventilation and the method applied do not provide the required level of air. Herein, the VENTSIM 5 simulation software will be used to assess the influence of many ventilation parameters in wind speed, which are of critical in practice [6].

For the development of the ventilation network, a main level of the ELMER 70D Mine Concession will be considered while considering the following as reference: Nv 200 – CX 786, Nv 300 – C 300.
In [7], the author refers to the appropriate design and implementation of mine ventilation systems, which are key to ensure mine worker safety. In this respect, the efficiency and reliability of mine ventilation systems depends on the numeric simulation of various regular and emergency situations in mines. Similarly, it implies the adequate development of mathematics and mine ventilation processes.

3. Contribution

3.1. Main data and calculation

The ELMER 70D Mine Concession is located 4,430 m above sea level; therefore, the air stream will increase 100% to 6 m3/min per person.

When hydrocarbon engine equipment is not used during operation, the total stream must be calculated in accordance with the formula presented below, before being compared against the explosive consumption stream. After obtaining each value, the highest value is determined as the total air requirement. Air demand inside the mine must be calculated under section d), article 252 of the regulation [8], considering the following formula:

$$QT0 = QT1 + QFu \quad (1)$$

Where: QT0 = Total stream for operation, GALLERY, LEVEL 100 and QT1 = The sum of air stream required by:

$$QFu = 15\% \text{ from } QT1 \quad (2)$$

In this sense, the following values must be previously determined.

3.1.1. Stream required by number of workers (QTr)

$$QTr = F \times N \ (m^3/min) \quad (3)$$

Where: QTr = total stream for "n" workers (m^3/min); F = Minimum stream per person according to range set forth in article 247 of the regulation; and N = Number of workers of largest crew on call.
Data: F = 6 m3/min (for an altitude of 4430 m above sea level); N = 20 collaborators.

Table 1. Stream required by the number of workers

Company	No of WORKERS ON CALL	DS 023-2017 EM (me/min)	Stream	
			M3/min	CFM
S.M.R.L. ELMER 70A	15	6	90	3,178
AVG Shaft Technologies S.A.	5	6	30	1,059
	QTr		120	4,237

3.1.2. Stream required by wood consumption (QMa)

$$QMa = T \times u \ (m^3//min) \quad (4)$$

Where:

QMa = stream required by tons of production (m3/min);

u = production factor
T = production in wet metric tons per crew on call.

Table 2. Production factor according to wood consumption

Zone	PRODUCTION TMH/GD	WOOD CONSUMPTION TN/GD	WOOD CONSUMPTION %	PRODUCTION FACTOR N M3/MIN	m3/min	CFM
	0	0	0.00%	0	0	0
	25.23	7.13	6.99%	0	0	0
	12.20	0	3.79%	0	0	0
		Qma			0	0

Data:

u = from 7.3%; therefore, it corresponds to 0 m3/min.; T = 37.53 TM/crew on call.

3.1.3. Stream required by temperature in work tasks (QTe)

$$QTe = Vm \times A \times N \ (m^3/min) \quad (5)$$

Where: QTe = stream on temperature (m^3/min); Vm = minimum speed; A = mean task area; and N = number of levels with temperature above 23°C, equal to a 30 m/min value, under range as per third paragraph of section d) in article 252 of the regulation.
Data: Mean temp. is 25°C; speed must be 30 m^3/min.; Vm = 30 m/min; A = Mean task area is 6.7 m2

Table 3. Minimum speed of airflow

Zone	Level with 24 to 29°C temperatures	Minimum speed (Vm) DS 023-2017 EM	Area (Aver.)	Stream	
				m3/min	CFM
CUERPOS	0	0	16.7	0	0
ELMER 70D, nivel 100	1	30	6.7	201	7,098
ESPERANZA	0	0	8.4	0	0
	QTr			201	7,098

3.1.4. Stream required by Explosive consumption (QEx)

$$QEx = A \times V \times N \ (m3/min) \quad (6)$$

Where: QEx = air stream required by consumption of detonated explosive (m3/min); A = mean task area (m2); V = minimum speed required by regulation (m/min); and N = number of blasting levels.
Data: QEx = 20 m/min (Air stream required for dynamite); A = 6.7 m2; V = 20 m/min (Air stream required for dynamite); N = 1; and QEx = 6.7 m2 × 20 m/min × 1 = 134 m3/min.

3.1.5. Stream required by leaks (QFu)

$$QFu = 15\% \times QT1 \ (m3/min) \quad (7)$$

Where:

$$QT1 = QTr + QTe + QMa + QEq \quad (8)$$

Table 4. Stream required by leaks

	m3/min	CFM
Q_{Tr}	120	4,237
Q_{Ma}	-	-
Q_{Te}	201	7,098
Q_{Ex}	134	4,732
QT1	455	16,067

Table 5. Stream required by total leaks

QT1 (m3/min)	Min. speed (Vm) DS 023-2017 EM	m3/min	CFM
455	15%	68	2.410
	QFu	68	2.410

Total stream of air requirement:

$$Q_{To} = Q_{T1} + Q_{Fu} \tag{9}$$

The following is required to analyze data and future design of ventilation system (Fig. 2)

Table 6. ELMER 70D air stream requirement

STREAM	m3/min	CFM
QT1	455	16.067
QFu	68	2.410
QT0	523	18.477

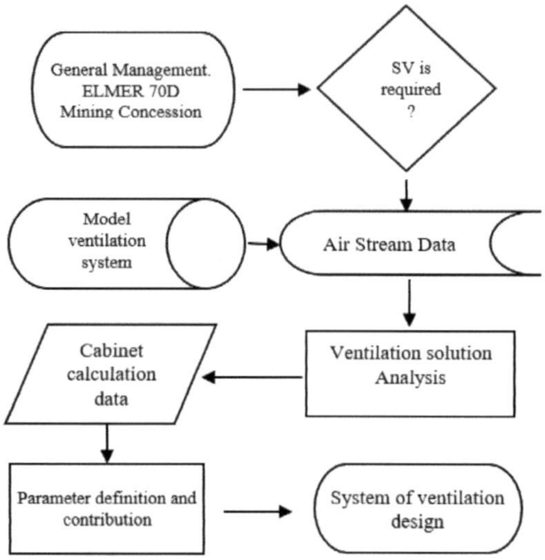

Figure 2. Design of ventilation system

3.2. Valuations of mining exploitation

Table 7. Total inlet of fresh air ELMER 70D

FRESH AIR INLET							
EST.	LEVEL	TASK	ZONE	DATE: 05/15/2019	TEMP. °C	AREA	Speed
				TIME		m2	m/min
EVPI-01	300	BM- 435	ESPERANZA	8:20:00 am	7,2	6,5	38
EVPI-02	100	BM- 435	ELMER 70D -100	9:12:00 am	7,3	6,7	25
EVPI-04	100	BM-ALEX 01	CUERPOS	10:21 am	7,2	7,2	28
STREAM		TOTAL INLET		532,99	18822		

Table 8. Total outlet of foul air ELMER 70D

FOUL AIR OUTLET									
EST.	LEVEL	TASK	ZONE	DATE: 05/22/2019	TEMP. °C	AREA	SPEED	STREAM	
				TIME		m2	n/min	m3/min	cfm
N-200	SURFACE	RB 593	CHIMENEA	9:21 am	10,2	4,5	12624	238,94	201.575
EVPS-02	SURFACE	RB 131	ELMER 70D 100	9:52 am	11,2	6,7	51372	307,04	184.564
					TOTAL OUTLET 545,98				19282

Table 9. Air covering

AIR COVERING		
	m3/min	cfm
Air inlet	532	18.822
Required stream	523	18.477
Covering (%)	110%	
Surplus (CFM)	345	

4. Results

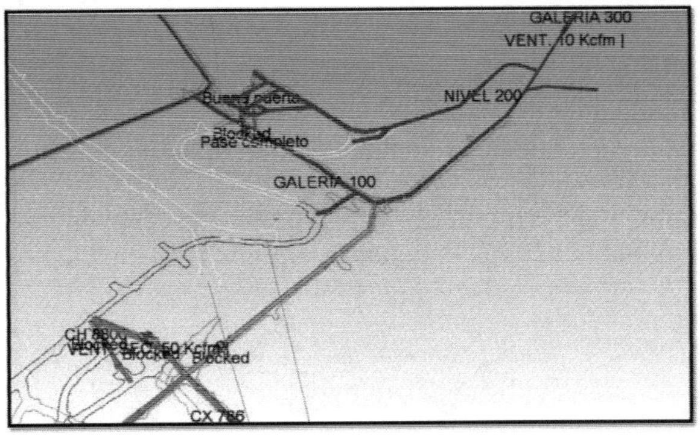

Figure 3. CM ELMER 70D 2019 simulation design

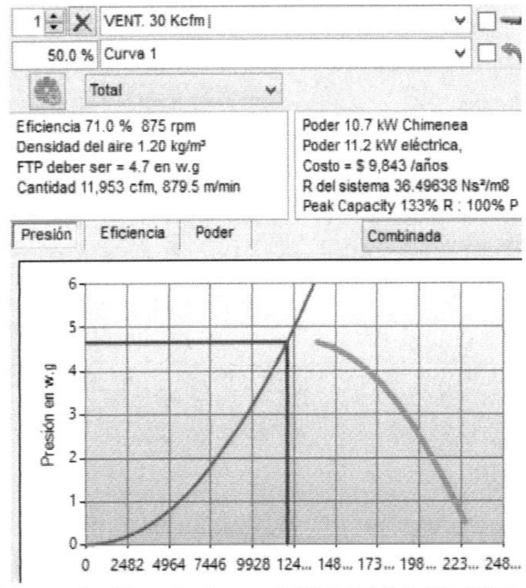

Figure 4. Simulation of CM ELMER 70D air stream

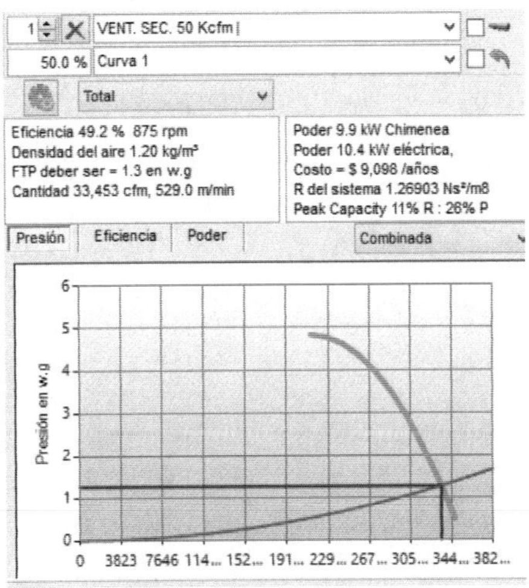

Figure 5. Simulation with N200 air stream upgrade.

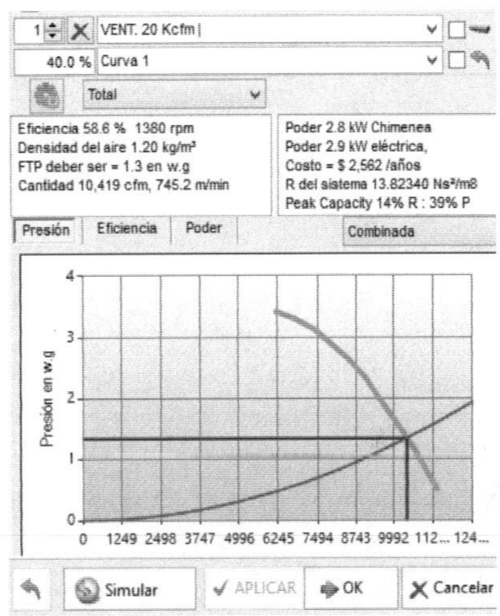

Figure 6. Simulation with N300 air stream upgrade

Level 100 has three chimneys. The first chimney is 200 m deep throughout the entire length of level 100, and it will be a new duct of fresh air inlet as shown in Fig. 3. The second chimney will be closed, no air will be blown into nor extracted from the chimney. The third chimney, which is 100 m from the second one, will be an outlet for foul air as it is connected to the surface.

These chimneys connect levels 100 and 300. There will be an extractor, a 20 kCFM primary fan, in the third chimney. Power results can be observed in Fig. 4.

Level 200 has a secondary fan with an efficiency at the intersection of limits, which is adequate, as shown in the following simulation results.

Level 300 has an auxiliary fan and 53-m long ventilation shafts with efficiency at the intersection of limits, which is adequate, as shown in the following simulation results. In Figures 5 and 6 an improvement in fan stream and power levels can be observed as they are at the corresponding intersection.

5. Conclusions

This study demonstrated the importance of chimneys, which will no longer be used as exhaust pipes but instead exclusively for ventilation. To this end, a main fan, at a capacity of 20 kCFM, will be installed to extract foul air from chimney No. 2

It was also concluded that when the diameter of ventilation shafts is small, there will be less air stream. Therefore, if mining operations expand in the ELMER 70D Mining Concession, a 0.35-m ventilation shaft will be required.

The ELMER 70D Mining Concession does not have a ventilation circuit. Instead, it features a natural draught system that does not allow good airflow. Then, planning a ventilation system for current activity and future advances. For example, both the simulation and cabinet calculations revealed that LEVEL 100 requires an airflow of at least 18477 CFM as the initial fresh air blown into the chimney at 10 kCFM was not effective enough for proper operation performance.

References

[1] Flores, V., Arauzo, L., Jara, J., Raymundo, C., "Optimized ventilation model to improve operations in polymetallic mines in peru", Smart Innovation, Systems and Technologies, 140, pp. 515-522.

[2] 24-2016-EM, D. (2016). Reglamento Seguridad y Salud Ocupacional en Minería DECRETO SUPREMO No 024-2016-EM. E.M., D.S. N° 024 -2016. Retrieved from http://www.osinergmin.gob.pe/seccion/centro_documental/PlantillaMarcoLegalBusqueda/DS-024-2016-EM.pdf 2.

[3] Pierola, O. (2015). Evaluación integral del sistema de ventilación, modelamiento y diseño mediante el uso del software ventsim visual avanzado en la U.M. Animon Cia. Minera Chungar. 208.

[4] De Villiers, D. J., Mathews, M. J., Maré, P., Kleingeld, M., & Arndt, D. (2019). Evaluating the impact of auxiliary fan practices on localised subsurface ventilation. International Journal of Mining Science and Technology, (xxxx), 2–10. https://doi.org/10.1016/j.ijmst.2019.02.008.

[5] Chang, P., Xu, G., Zhou, F., Mullins, B., Abishek, S., & Chalmers, D. (2019). Minimizing DPM pollution in an underground mine by optimizing auxiliary ventilation systems using CFD. Tunnelling and Underground Space Technology, 87(February), 112–121. https://doi.org/10.1016/j.tust.2019.02.014.

[6] Jin-Ming, M. O., She-Ji, Z., & Wei, M. (2019). Analysis on the influence of the diameter of the suction air inlet of the air duct on the wind flow field. IOP Conference Series: Earth and Environmental Science, 223(1), 0–6. https://doi.org/10.1088/1755-1315/223/1/012023.

[7] Semin, M. A., & Levin, L. Y. (2019). Stability of air flows in mine ventilation networks. Process Safety and Environmental Protection, 124, 167–171. https://doi.org/10.1016/j.psep.2019.02.0067

[8] Educación, M. De. (2016). Aprueban Reglamento de Seguridad y. 595392–595447.

IOP Conf. Series: Materials Science and Engineering **689** (2019) 012018 doi:10.1088/1757-899X/689/1/012018

Controlled Trim-Blasting Model to Improve Stability and Reduce Vibrations at a Production Gallery of the San Ignacio de Morococha S.A.A. Mining Company

C Camallanqui-Alborque[1], G Quispe[2], C Raymundo-Ibañez[3,*]

[1] Facultad de Ingeniería, Universidad Continental, Lima 15046, Perú.
[2] Ingeniería Industrial, Universidad Nacional Autónoma Altoandina de Tarma, Tarma 12651, Perú.
[3] Dirección de Investigación, Universidad Peruana de Ciencias Aplicadas (UPC), Lima 15023, Perú.

*carlos.raymundo@upc.edu.pe

Abstract. This paper presents a blasting method called controlled trim blasting, in which the rock mass of an unstable gallery where high levels of vibration have been detected is analyzed. This methodology comprises a drilling mesh with two-contour gallery assessment, producing its drilling machines and determining the type of explosive used and burden and spacing, which will be detonated after the internal blasting. Further, the internal blasting will possess its drilling machines, burden, spacing, and a second type of explosive. The separation of the gallery into smaller parts will improve the blasting, as verified in the recorded simulation. In addition, the rock-mass stability improves because the explosives used in the perimeter of the gallery are low-power with mild detonation pressure, which does not generate high levels of vibration. This is a practical and efficient method in areas where the rock mass is not good or there is a mixture of rock types.

1. Introduction
This document, has been written as a result of the deficiency generated by rock blasting in production galleries. Every time explosives are detonated in the gallery, the level of vibration generated affects the rock and its physical environment; therefore, it is necessary to provide greater support in the gallery, which leads to higher costs in the global production of the pit [1]. The intention is to use the trim-blasting method to reduce the blasting vibrations produced and generate greater stability in the gallery.

Therefore, a preliminary study is conducted on technical parameters such as the type of rock and its physical properties and dimensions of the gallery. Then, as per these results, a drilling mesh is constructed taking technical blasting considerations into account to correct the vibration levels by using either suitable explosives and an appropriately designed drilling mesh or the explosions delays necessary for blasting optimization and improving gallery stability, thus lowering mine support costs [2].

2. Theoretical Framework
The authors Bernaola Alonso, Castilla Gómez, & Herrera Herber [3], in their "Perforación y voladura de rocas en minería" (Drilling and blasting of rocks in mining) book, assert that a very important mine

exploitation factor is blasting and that it is closely related to the blowing techniques, drilling systems, controlled blasting, and the correct use of the explosives that are applied over the rock, thus leading to a generalized method of efficient drilling-mesh designs according to the type of rock mass, safety measures, and their influence on the rock during drilling and blasting.

To better understand this subject, the basic terms used in drilling, blasting, and geomechanics are described below, as define by Ames Lara & Ames Lara [4] in their article "Diseño de las mallas de perforación y voladura utilizando la energía producida por las mezclas explosivas" (Design of drilling and blasting meshes using the energy produced by explosive mixtures).

Blasting—The violent explosion in a gallery, front, chimney, cruise, etc. with assistance from explosives.

Density—The ratio of the volume and mass or of the mass of a substance and that of an equal volume of another substance that is considered a model.

Detonation—Rapid and sudden explosion caused by a relatively stable explosive.

Vibrations—It is the progressive increase of the waves that elastically and plastically affect the mass in solid or liquid masses on a body.

Stability—Natural state of equilibrium of a rock mass where the vertical tensions are equal to the horizontal ones, not causing rocks to detach or fall.

Gallery—Mining work, where mines are excavated horizontally or almost horizontally with different sizes and shapes, serving as access and a path throughout the facilities and as a road to extract the minerals obtained.

3. State of the Art

The purpose of this research work is to determine the influence of the vibrations produced by blasting rocks in a gallery and how these vibrations affect the stability of the walls and ceilings, often requiring support techniques. Therefore, we must first identify the recurring problem: rocks falling in the blasting areas. Second, we proceed to assess an improved mesh drilling and controlled blasting method. The document by Salum & Murthy [5] debates the determination of the optimal round lengths of excavation in various kinds of rocks and control over tunnel excavation. The excavation sequence and requirements for the top header and bank are also addressed.

In the blasting operation process according to Xie, Lu, Gu, & Wang [6], shock, stress , and seismic waves develop later. Owing to the existence of these waves, the internal structure of the rock mass is inevitably influenced. The vibration speed exceeding a critical value can cause the formation of internal cracks in the rock mass, potentially damaging the protected structure.

According to [7], when the blasting and explosive area is determined, the loading structure plays an essential role in blasting-energy transmission.

In the research by Ebadzadeh, Samareh, Eslami, Khoshrou, & Shahriar [8], from the first 11 controllable and uncontrollable parameters the influence of which on the speed of the seismic waves has been assessed by several researchers, the waves that match the propagation path corresponding to each of the 95 recorded in the seismic mapping on the geological and geometric model of the mine were selected.

In the research by Park, Kim, & Kwon [9], a sensitivity analysis of the parameters related to the reduction factor was conducted based on the intact rock; further, the properties in the soft explosion design show the variation of the factor of reduction. When increasing the constant value, the variation of the reduction factor was more significant.

4. Contribution

Considering the optimization and control of the explosions in the damaged zone of galleries to minimize the vibration and the damage in the rock and the design of an excavation method to reduce vibrations in difficult geological areas, the analysis of instability in underground gallery construction and the research study of the development of a damaged area and the influence of mechanical behavior in gallery construction.

The problem is the damage caused by the vibrations produced by rock blasting in the perimeter of the gallery. Considering this, the proposed model uses controlled trim blasting to minimize the influence of vibrations and yield greater stability in the hanging wall, footwall, and perimeter of the gallery to improve security and lower mine support costs. A controlled blasting uses low-power explosive charges placed with a very small distance between one another and from the edges of the gallery profile. These charges are fired simultaneously to avoid over-breakage and cracks, obtain smooth profiles, and above all, improve stability.

The activities performed according to the problem assessment and solution requirements are as follows:

- To identify the type of rock found in the gallery and its physical properties
- To determine the gallery area
- To identify the accessories required
- To collect drilling data such as drilling mesh, drilling diameters, drilling lengths, load, spacing, and use of relief bores
- To collect blasting data, such as the type of explosive used, and quantity of explosives
- To indicate the sequence of ignition, rock movement, power factor, and load factor
- To determine new data based on mine data that may help develop the controlled-blasting model
- To assess the controlled blasting model through simulation software

Table 1. Data Collected

Gallery Width	4.5	M
Gallery Height	4.5	M
Drilling Length	3.50	M
Actual Drilling Length	3.15	M
Rock type	Granite	Soft
Rock density	2.5	Tn/m^3
Production bore	50	mm
Relief bore	102	mm
Bieniawski classification	41	RMR Class III
Bores used	51	bores

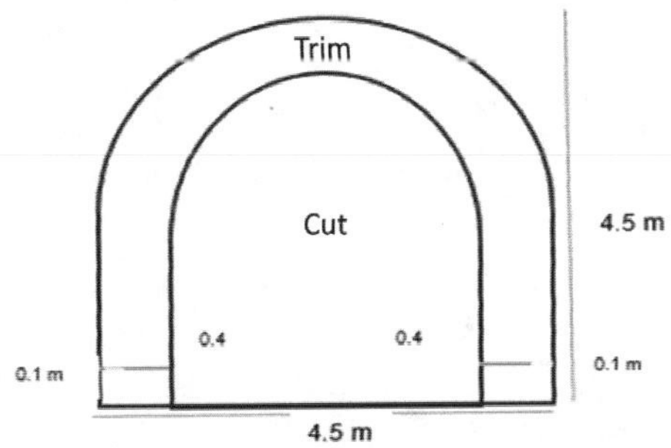

Figure 1. Define the cut and the trim of the gallery

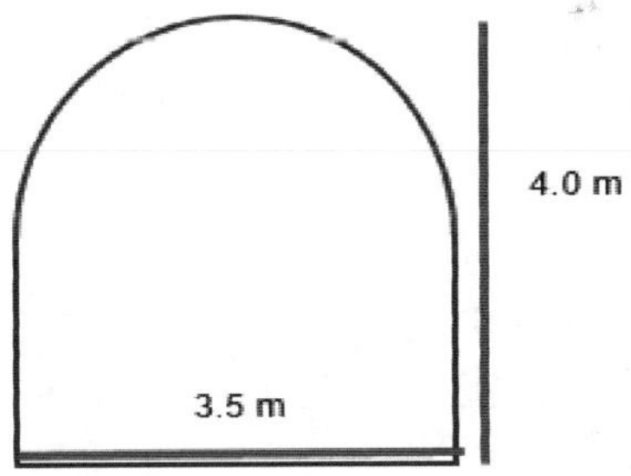

Figure 2. Gallery cut area

5. Validation

To validate the methodology, the data collected were taken and the calculations were performed. First, the trimming and internal areas of the gallery and their perimeter are defined.

Total area

$$S = \frac{BxH}{12} x(8+\pi); \quad S = \frac{4.5x4.5}{12} x(8+\pi); \quad \boldsymbol{S=18.801\ m} \tag{1}$$

Total perimeter

$$P = \sqrt{18.801}\, x\, 4\, ; \quad \boldsymbol{P = 17.34\ m} \tag{2}$$

Then, we determine the number of trimming bores using the following formula:

Number of Bores = (P – width new) / E

Number of Bores = (17.34 – 3.5) /0.31 = 44.64

IOP Conf. Series: Materials Science and Engineering **689** (2019) 012018 doi:10.1088/1757-899X/689/1/012018

Number of Bores = 45 bores

Then, we only consider the cutting area and proceed to calculate the area, perimeter, and number of bores.

$$S = \frac{BxH}{12}x(8+\pi); \quad S = \frac{3.5x4.0}{12}x(8+\pi); \quad \boldsymbol{S = 13.00\ m} \tag{3}$$

$$P = \sqrt{13.00}\ x\ 4; \quad \boldsymbol{P = 14.42\ m} \tag{4}$$

$$\textbf{Number of bores} = \frac{p}{d_t} + (CxS) = \frac{14.42}{0.62} + (1.5x13.00) = \boldsymbol{43\ bores} \tag{5}$$

Then, the load and spacing, amount of charge per bore, and type of explosive to be used are developed.

Table 2. Burden and spacing formula

Cut section	Burden value	Section side
First	B1=1,5 x D2	B1 x √2
Second	B2=B1 x √2	1,5 x B2 x √2
Third	B3=1,5 x B2 x √2	1,5 x B3 x √2
Fourth	B4=1,5 x B3 x √2	1,5 x B4 x √2

Table 3. Burden and spacing chart

BURDEN (m)	SPACING (m)
0.15	0.22
0.22	0.46
0.46	0.97
0.97	2.07

Theoretical bore length = 3.5 m; Actual drilling length = 3.15 m

Broken volume

V = Area × (Actual drilling length) = 13.00 × (3.15) = **40.95 m^3**

Load quantity

Lq = V × Factor = 40.95 × 1.3 = 53.23-kg. explosive per shot (Table 4)

Average bore load

Qp = Lq / No. = 53.23 / 40 = 1.33 kg/bore

Explosives:

Bores: Dynamite Semexsa 65, 1.5" (diameter) × 12" (length), 0.26 kg/cartridge (Table 5)

Table 4. Load quantity

Type of rock	Factor
Very difficult	1,5 to 1,8
Difficult	1,3 to 1,5
Easy	1,1 to 1,3
Very Easy	1,0 to 1,2

Table 5. Explosive charge distribution in all bores

Drill	N° bore	Load (kg)	Improved Load (kg)	Explosives per bore (calculation)	Cartridges per bore (improved)	#Total cartridges	TOTAL LOAD
Relief	3.00	0.00	0.00	0.00	0.00	0.00	0.00
Boot	3.00	1.33 x 1.50	2.00	5.42	5.50	16.50	5.99
Aids	17.00	1.33 x 1.20	1.60	4.34	4.50	76.50	27.15
Quarry	6.00	1.33 x 0.90	1.20	3.25	3.50	21.00	7.19
Upper	8.00	1.33 x 0.60	1.20	3.25	3.50	28.00	9.58
Pulling	6.00	1.33 x 1.00	1.33	3.62	4.00	24.00	7.98
Total No. of bores	43.00	Improved Total Load (kg)				166	57.89

Finally, after completing the entire calculation process for the cut of the internal gallery, we proceed to conduct the same process in the contour for controlled trim blasting.

Number of bores = 45 bores

Spacing = 0.31 m

Burden = 0.40 m

Trimming area = total area − new area = 18.80 − 13.00 = 5.80 m^2

Load factor = 1.3

Actual Length = 3.15

Volume = Area × Actual length = 5.80 × 3.15 = 18.27 m^3

Load quantity = Volume × load factor = 18.27 × 1.3 = 23.75 kg × shot

Load per hole = Load quantity / Number of bores = 23.75 / 45 = 0.53 kg/bore

ton = Volume × Density = 18.27 × 2.5 = 45.68 ton

Explosive: Exadit 45 weight 0.076 g − 328 per box

IOP Conf. Series: Materials Science and Engineering **689** (2019) 012018 doi:10.1088/1757-899X/689/1/012018

Number of cartridges = (Load per bore / weight) × number of bores = (0.53 / 0.076) × 45 = 313.8 cart.

In the simulation, the quantity of explosives will be considered not only in the cut but also in the trim area. In addition, the production and relief diameters, spacing, actual perforation length per bore, distribution of the bores in the drilling mesh and type of explosive used in each blasting will be required.

The number of bores in the determined mesh will be carried out using the simulation software.

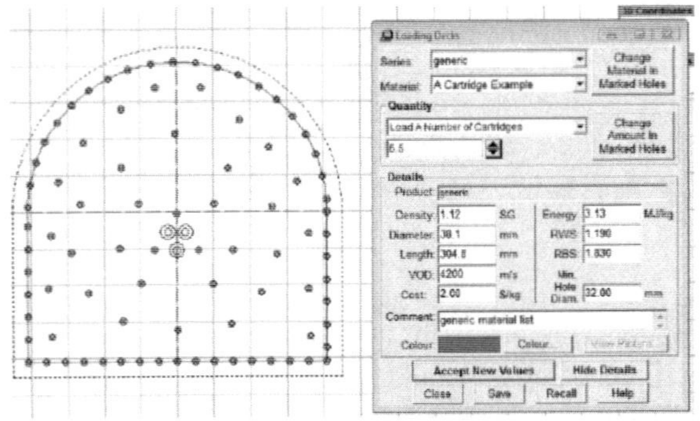

Figure 3. Bores in the pre-cut

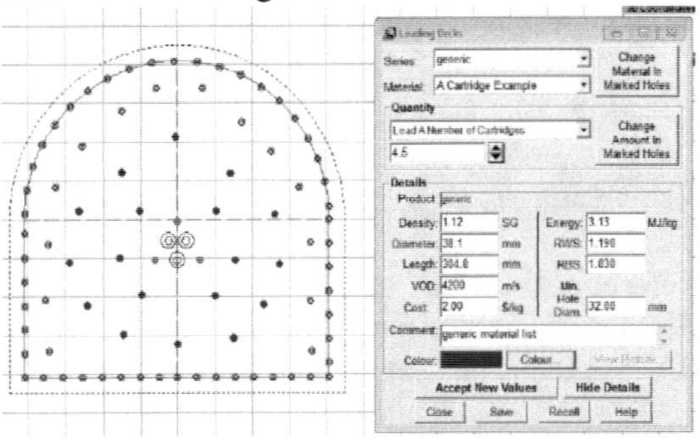

Figure 4. Relief bores

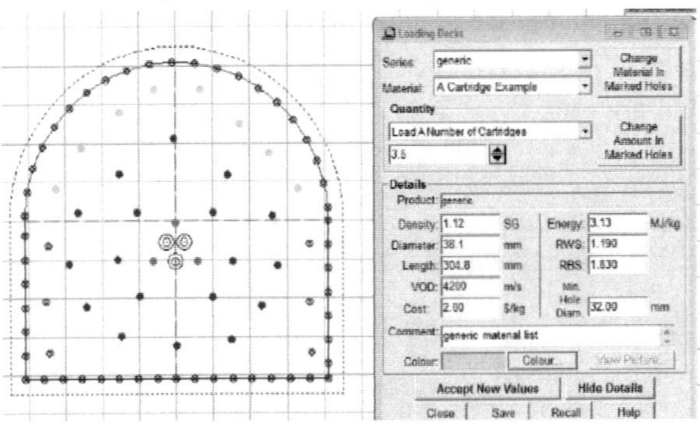

Figure 5. Upper Bores

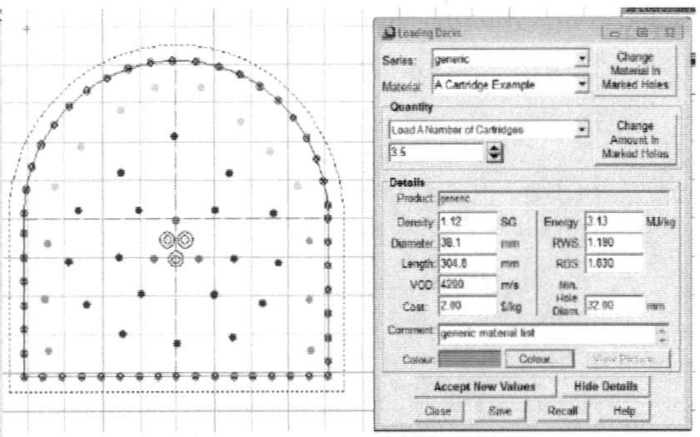

Figure 6. Contour bores

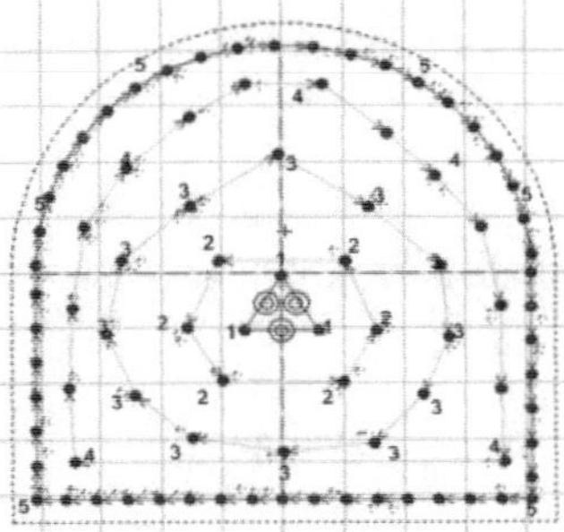

Figure 7. Ignition sequence of perforation mesh

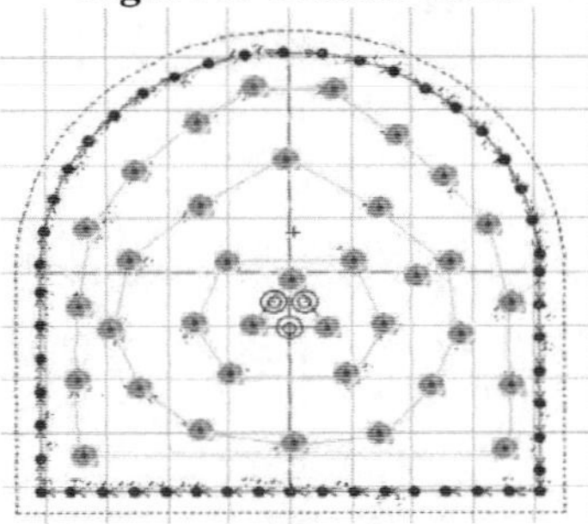

Figure 8. Detonation of bores

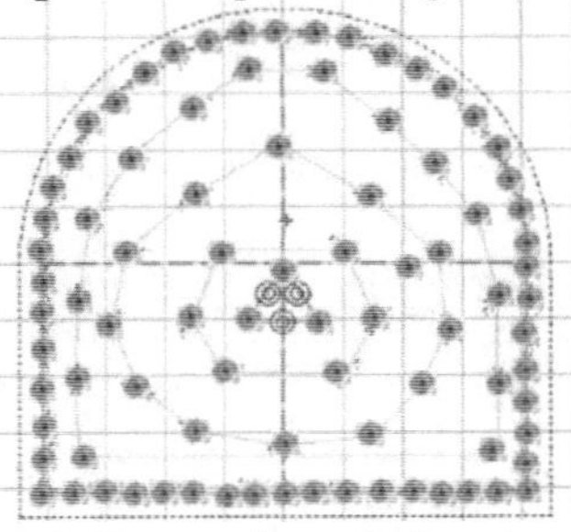

Figure 9. Detonation of trimming bores

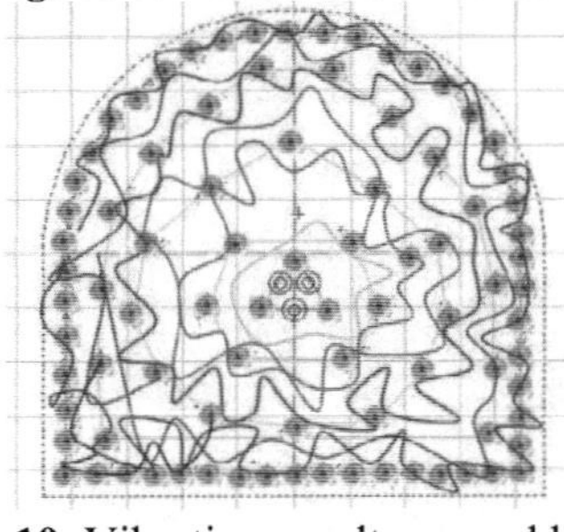

Figure 10. Vibration results caused by the blasting

6. Conclusions

- The controlled trim-blasting model was efficiently developed by reducing the vibrations in the hanging wall, footwall, and contours of the gallery using parameters that can be measured according to the characteristics that are available in the production gallery at the San Ignacio de Morococha SAA mining company.
- Some of the effects of the implementation are as follows: the working time improved because pneumatic bore and support work were conducted after the blasting; the reduction of the high costs involved in the making of support of any type; and the improvement in the drilling mesh in the production gallery.
- As shown in Figure 10, the contours have not been damaged, nor has breakage been caused, thus improving the stability of the rock mass while reducing the vibrations to a level that renders a stable gallery with little or almost no support.

References

[1] Rojas, K., Aramburú, V., Ramos, E., Raymundo, C., Moguerza, J.M., "Six sigma-based optimization model in hauling cut and fill exploitation activities to reduce downtime in underground mines in Peru", Advances in Intelligent Systems and Computing, 971, pp. 365-375.

[2] Caballero, E., Calixto, R., Arauzo, L., Raymundo, C., " Model for Optimization of Drilling, Blasting and Fragmentation Processes in medium mining", IMCIC 2019 - 10th International Multi-Conference on Complexity, Informatics and Cybernetics, Proceedings. 2, pp. 52-57.

[3] Bernaola Alonso, J., Castilla Gómez, J., & Herrera Herbert, J. (2013). Perforación y voladura de rocas en minería. Retrieved from http://oa.upm.es/21848/

[4] Ames Lara, V. A., & Ames Lara, V. A. (2008). Diseño de las mallas de perforación y voladura utilizando la energía producida por las mezclas explosivas. Universidad Nacional de Ingeniería. Retrieved from http://repositorio.uni.edu.pe/handle/uni/635

[5] Salum, A. H., & Murthy, V. M. S. R. (2019). Optimising blast pulls and controlling blast-induced excavation damage zone in tunnelling through varied rock classes. Tunnelling and Underground Space Technology. https://doi.org/10.1016/j.tust.2018.11.029

[6] Xie, L., Lu, W., Gu, J., & Wang, G. (2018). Excavation Method of Reducing Blasting Vibration in Complicated Geological Conditions. Shock and Vibration. https://doi.org/10.1155/2018/2518209

[7] Gu, W., Wang, Z., Chen, J., Liu, J., & Lu, M. (2015). Experimental and Theoretical Study on Influence of Different Charging Structures on Blasting Vibration Energy. Shock and Vibration. https://doi.org/10.1155/2015/248739

[8] Ebadzadeh, M. M., Samareh, H., Eslami, M., Khoshrou, S. H., & Shahriar, K. (2017). Optimization of a nonlinear model for predicting the ground vibration using the combinational particle swarm optimization-genetic algorithm. Journal of African Earth Sciences. https://doi.org/10.1016/j.jafrearsci.2017.04.029

[9] Park, S., Kim, J. S., & Kwon, S. (2018). Investigation of the development of an excavation damaged zone and its influence on the mechanical behaviors of a blasted tunnel. Geosystem Engineering, 21(3), 165–181. https://doi.org/10.1080/12269328.2018.1461139

Variation Propagation in Multistage Machining Processes Using Dual Quaternions

F Yacob[1] and D Semere[1]
[1]Production Engineering, Royal Institute of Technology, Brinellvägen 68, 11428 Stockholm, Sweden

E-mail: filmona@kth.se

Abstract. The application of rigid transformations matrices in variation propagation has a long tradition in manufacturing community. However, the matrix-based modeling of variation propagation in multistage machining processes is complicated. Moreover, there is a need to improve the computational efficiency of manipulation of geometrical models for variation analysis purposes. This paper introduces the representation of rigid transformation by dual quaternions, which have a computational advantage and mathematical elegance compared to matrices. In comparison to a commercial tool, the implementation of the purposed method predicted parallelism with an average error of 4.3 % in a hypothetical two stage machining process. The proposed approach has a potential to be an alternative to matrix based rigid transformation practices in variation and tolerance analysis.

1. Introduction

Variation in manufactured parts is an expected characteristic due to variation in machining processes. The variation can be caused by many sources, including loose locator, wear, datum variation from upstream, and a tooltip deviation from its nominal path. The combined effect of these variation sources makes it non-trivial task to predict the resulting geometry of a part, and thereby harder to entertain what-if scenarios.

The existing methods of variation propagation methods apply matrices to capture the effect of the sources of error. One of the common methods, Stream of Variation, utilizes matrix representation of the rotational and translational transformation of geometrical models [1]. Such approaches become mathematically more complex when some of the assumptions relaxed. For instance, the position of the secondary locators is often considered at the same height. The generic setup model of parts and fixtures is very complicated, often reduced to orthogonal assembly cases [2].

The use of Skin Models Shapes has enabled inclusions of form errors in variation propagation related analysis [3]. Skin Model shapes have been applied in tolerance analysis [4], variation analysis [5], [6], and anomaly detection [7]. The underlying principle of such analysis is an assembly of Skin Model Shapes with help of transformation matrices. However, depending on the density of points per model and number of models, computational efficiency suffers on performing operations such as registration [5]. Since the continuous registration steps are required, any improvement in computational efficiency is desirable.

In light of these, quaternions and their extensions dual quaternions, have a potential in variation propagation analysis without the dependency on transformation matrices. Dual quaternions perform better in terms of computational efficiency, speed and robustness compared to matrices [8], [9]. Moreover, dual quaternions represent rotation and translation in a compact unified single variable of 8 dimensions. Rotation is computed based on axis of rotation, not constrained to the series of orthogonal

IOP Conf. Series: Materials Science and Engineering **689** (2019) 012019 doi:10.1088/1757-899X/689/1/012019

rotations, which makes it much easier to iteratively optimize over [10]. Dual quaternions have been successfully applied in object blending and skinning in computer graphics [11], hand-eye calibration in robotics [12] and in kinematics of industrial robots [9], [13] and pair-wise point cloud registration [14], to mention a few.

This paper introduces dual quaternions for the first time in performing variation propagation in multistage machining. Using the dual quaternions, the angle and the axis of rotation of the deviated assembly feature from corresponding locators' plane is first computed for registration purpose. Following this, the distance between the feature's and locators' planes is used to obtain the translational part of dual quaternions. Once the registration is completed, the machining variation is induced by varying the depth cut. These steps are repeated at each station to predict the product geometry in subsequent stages.

2. Dual quaternions

Quaternions are compact mathematical representation of orientation of an object in space [15]. The fundamental concept is that 4 dimensional properties are required to represent 3D rotaion. The quaternions has two parts, the real part q_0 and the vector part $\bar{q}=q_1 i + q_2 j + q_3 k$, which can be denoted by

$$\tilde{\mathbf{q}} = q_0 + q_1 i + q_2 j + q_3 k = (q_0, \bar{q}) \tag{1}$$

where q_0, q_1, q_2 and q_3 quaternion properties of the unit vectors and i, j, and k, such that $i^2 = j^2 = k^2 = ijk = -1$.

The quaternion properties are computed using angle θ and axis of rotation $\mathbf{n}$ between an object and reference object.

$$\tilde{\mathbf{q}} = \left(\cos\left(\frac{\theta}{2}\right), \mathbf{n}\sin\left(\frac{\theta}{2}\right)\right) = (q_0, \bar{q}) \tag{2}$$

Dual quaternions, as extension of quaternions, were derived from the field of dual algebra [16]. Dual quaternions embed the rotational and translation aspect of rigid motion simultaneously. Dual quaternions have two part: the real component $\tilde{\mathbf{q}}_r$ and dual components $\tilde{\mathbf{q}}_d$, which can be denoted by

$$\hat{\mathbf{q}} = \tilde{\mathbf{q}}_r + \epsilon\tilde{\mathbf{q}}_d \tag{3}$$

where ϵ is a dual-operator such that $\epsilon \neq 0$, $\epsilon^2 = 0$.

Equation (3) is equivalent to the homogenous transformation matrix of 4 by 4.

3. Assembly and feature representation

3.1 Assembling part to a fixture

In multistage machining, variation propagates due to use of machined feature as an assembly feature in succeeding stations. To assemble a part to a 3-2-1 layout fixture, for example, specific features should be registered to their corresponding locators one at a time. For planar surface, the normals to the planes of the locators and features can be used to determine the necessary angle and axis of rotation required to assemble the geometric model to fixture. Mathematically, for an angle θ between two of the planes of two point clouds whose normal $\mathbf{n}_a \in \mathbb{R}^3$ and $\mathbf{n}_b \in \mathbb{R}^3$, the axis of rotation $\mathbf{n} \in \mathbb{R}^3$ is

$$\mathbf{n} = \mathbf{n}_a \times \mathbf{n}_b \tag{4}$$

The dual quaternion representation of pure rotation $\hat{\mathbf{R}}$ and pure translation $\hat{\mathbf{T}}$ are denoted by

$$\hat{\mathbf{R}} = \left(\cos\left(\frac{\theta}{2}\right), \mathbf{n}\sin\left(\frac{\theta}{2}\right)\right) + \epsilon(0_{1X4}) \tag{5}$$

$$\hat{\mathbf{T}} = ([1, 0, 0, 0\,]) + \epsilon(0, \frac{\bar{t}}{2}) \tag{6}$$

where $\bar{t} \in \mathbb{R}^3$ is the vector part of the quaternion representing the distance between corresponding points of the two planes. For rigid translation, the $\bar{t}$ is the same for all points.

The combined transformation, rotation followed by a translation, becomes the product of the two dual quaternions.

$$\hat{\mathbf{q}} = \hat{\mathbf{T}} \times \hat{\mathbf{R}} \quad (7)$$

For a 3-2-1 fixture layout, a part is assembled in a sequence of primary, secondary and tertiary locators. To assemble the part model to secondary and tertiary locators, the model is translated in the direction where primary and tertiary planes, and primary and secondary planes, intersect, respectively. Thus, dual quaternion representation of the total transformation $\hat{\mathbf{A}}_{\text{total}}$ becomes

$$\hat{\mathbf{A}}_{\text{total}} = \hat{\mathbf{T}}_{\text{t}} \times (\hat{\mathbf{T}}_{\text{s}} \times \hat{\mathbf{R}}_{\text{s}}) \times (\hat{\mathbf{T}}_{\text{p}} \times \hat{\mathbf{R}}_{\text{p}}) \quad (8)$$

where the subscripts p, s and t denote the primary, secondary and tertiary datum/ locators, respectively.

Following the definition transformation of point by dual quaternions, the point cloud of a feature with v points $\mathbf{A} = \{a_i \in \mathbb{R}^3 : i = 1 \ldots v\}$ is transformed to new position $\mathbf{A}'$ by applying

$$\mathbf{A}' = \hat{\mathbf{A}}_{\text{total}} \mathbf{A} \, \hat{\mathbf{A}}^{*}_{\text{total}} \quad (9)$$

3.2 Representation of machined surface

The dual quaternion representation of machined surface is based on the angular and translational distance between an actual and a nominal machined surface. These deviations can be computed by utilizing the landmarks (dummy) points as reference points that are not affected by the machining processes. The landmarks are easier to locate when placed at some distance from the point clouds of the model.

Without loss of generality, a 2D example of machining setup shown in Figure 1, the machined surface, which can be represented by coplanar points, remains horizontal despite the locator deviations. However, when the part is moved out of the setup by coinciding the landmarks, the machined surface results in orientation and position errors. Figure 1 shows an illustration the resulting machined surfaces before and after landmarks are coincided.

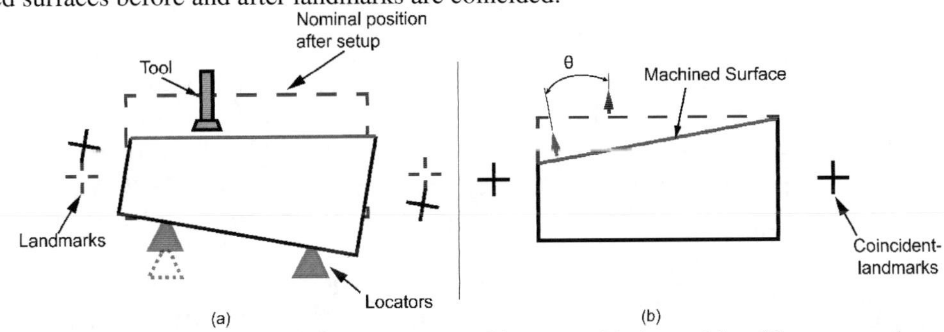

Figure. 1 An illustration of effect of deviation of locators. (a) Assembly of locators and part, (b) effect of coinciding landmarks

Mathematically, the normal of the displaced landmarks' plane $\mathbf{n}_{\text{m}}$ and that of the refence landmarks $\mathbf{n}_{\text{n}}$, the axis of rotation is computed from $\mathbf{n}_{\text{nm}} = \mathbf{n}_{\text{n}} \times \mathbf{n}_{\text{m}}$. The dual quaternion representation of the machined feature $\mathbf{q}_{\text{nm}}$, derived from pure rotation $\hat{\mathbf{R}}_{\text{nm}}$ and pure translation $\hat{\mathbf{T}}_{\text{nm}}$ in the direction of $\mathbf{n}_{\text{n}}$ becomes

$$\hat{\mathbf{q}}_{\text{nm}} = \hat{\mathbf{T}}_{\text{nm}} \times \hat{\mathbf{R}}_{\text{nm}} \quad (10)$$

The transformation of the nominal feature $\mathbf{N} = \{n_i \in \mathbb{R}^3 : i = 1 \ldots w\}$ by $\hat{\mathbf{q}}_{\text{nm}}$ gives the point cloud of the machined feature $\mathbf{M}'$ with w points.

$$\mathbf{M}' = \hat{\mathbf{q}}_{\text{nm}} \mathbf{N} \, \hat{\mathbf{q}}^{*}_{\text{nm}} \quad (11)$$

In line with Skin Model Shape operations [4], the point clouds of the model and machined surface, obtained from Eq. 9 and Eq. 11 give

$$\mathbf{A}_{\text{m}} = \{(\mathbf{A}', \mathbf{M}') \in \mathbb{R}^3\} \quad (12)$$

IOP Conf. Series: Materials Science and Engineering **689** (2019) 012019 doi:10.1088/1757-899X/689/1/012019

3.3 Setup in multistage processes

In multistage machining, the machined feature becomes an assembly feature in succeeding stations. The datum feature in the station is assembled to the fixture of the next station. Thus, the shape model is first rotated 180°, say around x-axis, which can be achieved by multiplying with the dual quaternion $\widehat{\mathbf{R}}_{\mathbf{A}_m}$

$$\widehat{\mathbf{R}}_{\mathrm{A_m}} = \left(\cos\left(\frac{180}{2}\right), [1, 0, 0]\sin\left(\frac{180}{2}\right)\right) + \epsilon(0_{1\times4}) \tag{13}$$

$$\mathbf{A}_{\mathrm{m}}^{'} = \widehat{\mathbf{R}}_{\mathrm{A_m}} \mathbf{A}_{\mathrm{m}} \widehat{\mathbf{R}}_{\mathrm{A_m}}^{*} \tag{14}$$

Following the model transformation by Eq. 14, the same steps of assembling the model to locators using Eq. 8 and Eq. 9. are repeated at each station. The resulting machined surface becomes the subject of analysis.

4. Demonstration case

To demonstrate the approach, a hypothetical 2 stage machining process with 3-2-1 fixture layout was set as shown in Figure 2. In station 1 and 2, feature A and B were represented by coplanar points set at different depth of cuts per experiment, respectively.

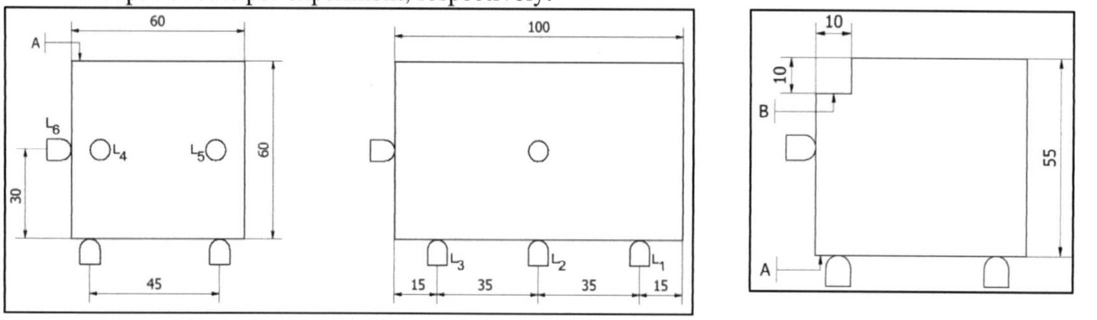

Figure 2. Part-fixture assembly. (a) side and front view of station 1(b) side view of station 2

Table 1. Inputs and predicted parallelism and normal of machined surface of station 2, based on nominal model and deviations of (0, 0, 0.5, 0, 0.5, 0, -0.1) in station 1.

Deviations (mm) [a]	CAM-Simulation (mm)[b]	Dual quaternions (mm)[b]
(0.3, 0, 0, -0.4, 0, 0, 0.6)	0.464, (0.0043, 0.0035, 0.9922)	0.468, (0.0043, 0.0034, 0.9923)
(0, -0.1, 0, 0, -0.2, 0, 0)	0.023, (0.0000, 0.0022, 0.9978)	0.024, (0.0000, 0.0023, 0.9977)
(0, 0, -0.2, 0, 0, 0, -0.4)	0.307, (0.0028, -0.0022, 0.9950)	0.307, (0.0028, -0.0023, 0.9948)
(0, 0.2, 0, 0, -0.3, 0.4, 0.2)	0.044, (0.0000, -0.0041, 0.9959)	0.051, (0.0000, -0.0046, 0.9953)
(-0.3, 0, 0.1, -0.2, 0, 0.2, -0.5)	0.593, (-0.0057 -0.0022 0.9921)	0.595, (-0.0057, -0.0023, 0.9920)

[a](Locators$_{1\text{-}6}$, Tooltip), [b]Parallelism of feature B relative to A, (normal of Feature B)

In each experiment, random locator deviations normal to fixtures were added that affect the length of the locators. With the help of Eq. 10, 3 dummy landmarks points were used transform the machined part from the fixture to the reference position. The experiment has been performed by inducing deviations to the setups. The prediction result of the second station, following the method proposed in this paper and the one acquired via Autodesk Inventor HSM CAM simulation 2016, is shown in Table 1. To performing machining simulation, this paper utilized the dual-quaternions' operations implementation in Matlab reported in [17].

One of the weakness of dual quaternions is that the direction of angle of rotation is indeterministic. Thus, the rotation angle should be subtracted from 180 degree when the angle is more than 90 degree. Moreover, the approach proposed in this paper applies the sequence of rotation followed by translation. However, the approach can be extended to a method to that simultaneously applies rotation and translation steps.

IOP Conf. Series: Materials Science and Engineering **689** (2019) 012019 doi:10.1088/1757-899X/689/1/012019

5. Conclusion

Variation propagation modelling and analysis in multistage machining based on matrices is non-trivial task. As an alternative, this paper applied dual quaternions in variation prorogation in multistage machining processes for the first time. The approach reduces the number of parameters in transformation matrices, which is an important step towards reducing the computational cost of manipulation of digital models. Towards this direction, this paper showed how variation propagation can be modelled using Skin Model Shapes and dual quaternions. The dual quaternions representation of orientation and position of a model was derived from the angles and positions between corresponding assembly features and fixtures. Following this, virtual machining was performed by positioning a point cloud on desired depth of cut per specific setup. The implementation of variation prediction based on dual quaternions provided parallelism accuracy of 4.3 % error compared to the one achieved by a commercial tool. In future, the approach will include the assembly of models with form errors using dual quaternions, and improvement to the overall accuracy of the predicted geometry.

Reference

[1] Abellán-nebot J V, Romero F, Julio S, and Mira S 2013 Manufacturing variation models in multi-station machining systems, *Int. J. Adv. Manuf. Technol.* pp 63–83

[2] Shi J 2006 *Stream of Variation Modeling and Analysis for Multistage Manufacturing Processes* CRC press

[3] Schleich B, Anwer N, Mathieu L, and Wartzack S 2014 Skin Model Shapes: A new paradigm shift for geometric variations modelling in mechanical engineering *CAD Comput. Aided Des.* **50** pp 1–15

[4] Schleich B and Wartzack S 2016 A Quantitative Comparison of Tolerance Analysis Approaches for Rigid Mechanical Assemblies, *Procedia CIRP* **43** pp 172–77

[5] Yacob F, Semere D, and Nordgren E 2018 Octree-Based Generation and Variation Analysis of Skin Model Shapes, *J. Manuf. Mater. Process.* **2** p 52

[6] Semere D, Yacob F, Hedlind M, and Bagge M 2018 Skin Model Based Tolerance and Variation Analysis, in *Skin Model Based Tolerance and Variations Analysis* pp 726-31

[7] Yacob F, Semere D, and Nordgren E 2019 Anomaly detection in Skin Model Shapes using machine learning classifiers *Int. J. Adv. Manuf. Technol.*

[8] Wang X and Zhu H 2014 On the Comparisons of Unit Dual Quaternion and Homogeneous Transformation Matrix **24** pp 213–229

[9] Sariyildiz E, Cakiray E, and Temeltas H 2011 A comparative study of three inverse kinematic methods of serial industrial robot manipulators in the screw theory framework *Int. J. Adv. Robot. Syst.* **8** pp 9–24

[10] Kenwright B 2012 A Beginners Guide to Dual-Quaternions **2679** pp 190–194.

[11] Kavan L, Collins S, Žára J, and O'Sullivan C 2007 Skinning with dual quaternions *Proc. symp. on Interactive 3D graphics and games* pp 39-46

[12] Daniilidis K 1999 Hand-Eye Calibration Using Dual Quaternions *Int. J. Rob. Res.* **18** pp 286–98

[13] Sariyildiz E and Temeltas H 2009 Solution of inverse kinematic problem for serial robot using dual quaterninons and plücker coordinates *IEEE/ASME Int. Conf. Adv. Intell. Mechatronics* pp 338–43

[14] Wang Y, Wang Y, Wu K, Yang H, and Zhang H 2014 A dual quaternion-based, closed-form pairwise registration algorithm for point clouds *ISPRS J. Photogramm. Remote Sens.* pp 63–9

[15] Hamilton W R 1866 *Elements of quaternions*. Longmans, Green, & Company

[16] Clifford W K 1882 *Mathematical papers*

[17] Leclercq G, Lefèvre P, and Blohm G 2013 3D kinematics using dual quaternions: theory and applications in neuroscience *Front. Behav. Neurosci.* **7** pp 1–25

IOP Conf. Series: Materials Science and Engineering **689** (2019) 012020 doi:10.1088/1757-899X/689/1/012020

Applying Lean Manufacturing Principles to reduce waste and improve process in a manufacturer: A research study in Peru

J Kaneku-Orbegozo[1], J Martinez-Palomino1, F Sotelo-Raffo[1], E Ramos-Palomino[1,*]

[1] Ingeniería Industrial, Universidad Peruana de Ciencias Aplicadas (UPC), Lima 15023, Perú.

*edgar.ramos@upc.edu.pe

Abstract. In the present scenario, lean manufacturing techniques are the most successful improvement concepts that many companies can apply in order to eliminate waste and non-value-added activities related to the manufacturing process. This paper discusses the implementation of lean manufacturing tools and an integrated framework between production planning and quality in a SME dedicated at a production of kitchen equipment. Studies of the company processes showed that the cutting and bending processes tend to create the most amount of sheet metal waste and non-value-added activities. The main objectives of the research were to standardize work, reduce waste, eliminate machine failures and develop the guidelines for a correct planning that meets quality requirements. The improvement results have shown a positive impact reducing the waste ratio and this has meant a saving of 13% in the manufacturing cost.

1. Introduction

Lean manufacturing is a methodology that reduces waste and increases productivity within the organization. It seeks to use fewer resources, such as manufacturing time, the number of workers and to develop highly efficient processes [1]. Likewise, waste reduction focuses on reducing overproduction, the amount of inventory of finished parts, defective products, waste, work in process inventory, waiting times and unnecessary movements [2]. However, the main objective of Lean Manufacturing is to maximize the value for the customer by eliminating the waste of production. The value is described as the ability to provide products at the right time and at the right price to meet the customer's needs. [3].

In this context, production waste can be eliminated through the development of different lean tools such as 5S, preventive maintenance, Kaizen, value flow mapping and standardization of work that can be applied in small-scale industries [4]. This interaction between the different tools encourages the reduction of reprocessing, defective products, high lead times and scrap, so the processes become smoother and reduce manufacturing costs [5]. The practice of the 5s principles is useful to improve the workplace by organizing, classifying, cleaning, standardizing and maintaining the different work stations in order to produce better quality products without the generation of waste associated with the work environment [6]. This Lean tool allows you to reduce quality errors and waste that occur during the production process. In this way, productive operations will generate homogeneous products [7].

Finally, preventive maintenance involves all employees of the organization in a constant equipment improvement. This kind of maintenance consists of a range of methods, which are known for being effective in improving reliability, quality, and production with simple maintenance actions [2].

IOP Conf. Series: Materials Science and Engineering **689** (2019) 012020 doi:10.1088/1757-899X/689/1/012020

The present investigation was undertaken at a company focused on the production of kitchen equipment in Lima, Perú. The main objectives were to present an overview of the manufacturing plant like the productive process and apply the lean manufacturing methodology in order to reduce waste since it is the main problem in the production line, generating an over-cost of 13% at the cost of manufacturing [3][8]. For this purpose, the operations in the workstations were standardized, the activities that did not generate added value were eliminated, the failure rate of the machinery was reduced and the work environment was organized.

2. Methodology

The implementation of lean manufacturing tools in this investigation was done step by step without losing sight of the coupling they must have with each other. For this reason, it was schematized as follows (fig.1):

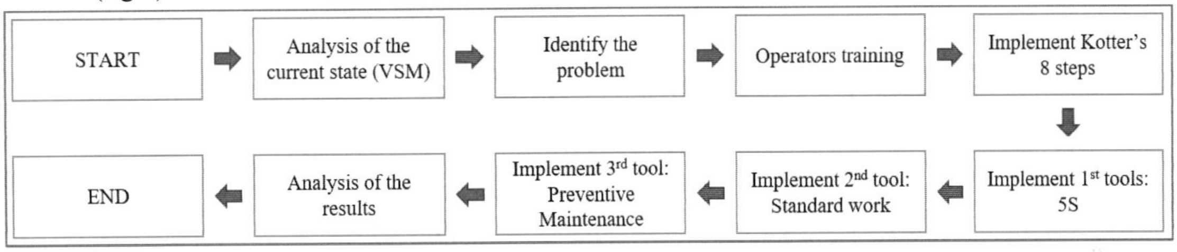

Figure 1. Implementation Methodology.

3. Literature Review

This section provides a review of lean implementation focused on 5S, Standard Work and Preventive Maintenance.

A review paper emphasizes the 5S implementation in the creation of a culture of safe, clean and organized company that is in continuous improvement based on risk assessment and cost for SMEs in India. At this way, the objective is to give a panoramic vision of the small business to improve its productivity to grow as company [6][9]. Furthermore, in another paper the implementation of 5S require the assignment of specific responsibilities for some members of each workstation and create forms for the standardization to use as a step by step manual [10] [5].

On the other hand, a furniture research paper shows the steps for the standard work since the elimination of the non-value-added activities to the calculation of the working ratio and workload fulfilling customer orders on time [11]. For this paper, one of the most important factors to take care is the use of the raw material because of the high costs they represent; for that reason, it was elaborated a process flow diagram to calculate the necessary resources for each area and not to waste raw materials [8].

Another research survey demonstrates the benefits of integrate and stablish a midpoint between production and maintenance planning as reducing the production costs and obtain the best solution for the company in order to minimize inventory levels and reduce workforce and equipment to produce just the necessary quantity with the major quality using the required resources [9]. Considering the scenario of small-scale industry, it is important to integrate philosophies that can work together to obtain better results. Total Productive Maintenance, Total Quality Management and operational performance is an example of these performances that bring a better plan of maintenance requirements having into consider state of final products.

The synergy of multiple techniques will make possible the production system improvement if they are correctly used. This does not mean implementation of one technique do not reduce wastes or improve productivity, but it is also focused in just one operation or activity and will not be expand to the entire company [12].

4. Results

The company ABC is a small-medium manufacturer of gastronomic equipment used in restaurants and fast food retails. The main products for these sectors are worktables, shelves, kitchen sinks, gas cookers and refrigerated tables. However, the first three products represent the highest volume of sales within the organization and their manufacturing process is similar from each other. Due to this reason, it will be considered as a single family to carry out the investigation. It should also be noted that the company manufactures the products according to the specifications of the customers, which is why it follows a "Make to Order" production.

The production process of ABC manufacturer is separated into five operations: design, cutting, bending, welding and surface finishing process. Despite the importance of all operations, the present research will only take into consideration the design, cutting and bending processes due to their relevance in the generation of waste during the manufacture of the products. These operations will be briefly detailed below:

- **Design process:** Products that will be fabricated using AutoCAD software are designed following the specifications granted by the client.
- **Cutting process:** The pieces to be cut are distributed in the steel plate according to the criteria of the operator and then they are cut.
- **Bending process:** In the hydraulic bending machine, the bends are made and the pieces are sent to the welding area.
- **Welding process:** The union of the pieces is done by TIG type welding, guaranteeing a better surface finish.
- **Surface finishing process:** In the assembled product, the weld cord, sharp edges are worked and the surface of the final product is polished to ensure the safety of the food.

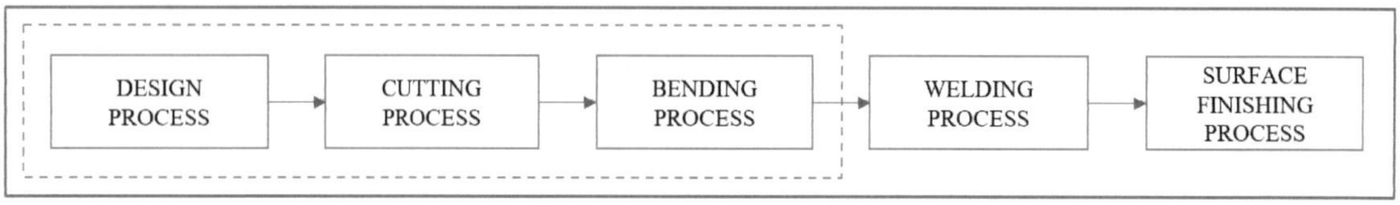

Figure 2. Production process.

The present research will focus on the design, cutting and bending process due to the costs related to the generation of waste in these operations. For this reason, to understand the activities of the processes to be study, the documentation of the operations, the productive areas performance and the manufacturing process were reviewed. In addition, the standard time for each activity was determined through a time study. In addition, the non-value-added activities were identified with the purpose of being deleted [4][11]. In addition, the lead-time of the raw material and the delivery of finished products to the customers were registered. In order to map the whole production process, the operations are present in a VSM diagram, showing their cycle time and the metrics related with the VSM [5][13]. Likewise, the WIPs are detailed between the areas and the Kaizen opportunities were identified [5].

The TAKT time was calculated through the customer demand and the available working time [9]. The current TAKT time is 4.4 hours, which is under the production time of 5.9 hours. This signifies that the organization is not fabricating the products fast enough to fulfill with the demand requested by customers. On the other hand, between the bending and welding area is a relevant amount of WIP relate to the high cycle time in the welding process. In addition, improvement opportunities were detected such as the work standardization, 5'S tool implementation and the development of preventive maintenance. This Kaizen opportunities are related to the variability in the work method at the cutting area, the frequently faults in the bending machine and the disorganized workplaces in the manufacturing line.

5'S Methodology: The five stages of the 5S methodology (Seiri, Seiton, Seiso, Seiketsu and Shitsuke) were implemented in the productive areas of the organization. Opportunities for improvement were identified and efforts were focused on them through the 5'S tool in order to reduce errors in production and manufacturing times. For this reason, unnecessary materials were discarded by identification cards and work tools were organized according to the scheme proposed by [6],[14]. In addition, a cleaning instruction was developed so that operators can carry out cleaning activities in a

IOP Conf. Series: Materials Science and Engineering **689** (2019) 012020 doi:10.1088/1757-899X/689/1/012020

standardized manner. Likewise, internal audits were carried out based on the guidelines presented by [15]. The joint work of the implementation of the 5S methodology and the standardization of processes resulted in a reduction in manufacturing times in all areas, as shown below.

Table 1. Results by area

PRODUCTION AREAS	TIME BEFORE IMPLEMENTATION	TIME AFTER IMPLEMETATION	TIME REDUCTION
Design	252.4 min	198.18 min	52.22 min
Cutting	59.24 min	50.28 min	8.96 min
Bending	32.58 min	26.51 min	5.3 min
Welding	190.71 min	132.50 min	58.21 min
Surface Finishing	71.65 min	58.87 min	12.78 min

Work Standardization: The work standardization activities were carried out following the guidelines proposed by for the areas of design, cutting and bending [16]. To determine the activities that add value, the value-added analysis matrix was used and the standard time for each activity was determined through a time study. After the standardization activity's in the process of the organization, the time of the cycle was reduced from 5.9 to 5.0 hours, so that the customer's demand would be met. For this, the standard operating procedure was drafted and validated in the areas of design, cutting and bending. In the design process, a new blueprint layout was developed where the pieces to be cut are distributed, in this way the workers reduce the waste index by cutting the pieces following a standard working method.

On the other hand, the procedures for cutting and bending operations were structured to reduce manufacturing time and use only activities that add value to the product. When comparing the waste index, a reduction of 6% was noticed after the implementation of the new procedures focused on determining the exact cutting area by means of the use of blueprints with precise measurements. This decrease in the percentage of waste has a positive impact on the reduction of manufacturing costs (13%) because stainless steel plates account for 80% of the costs associated with the use of raw materials.

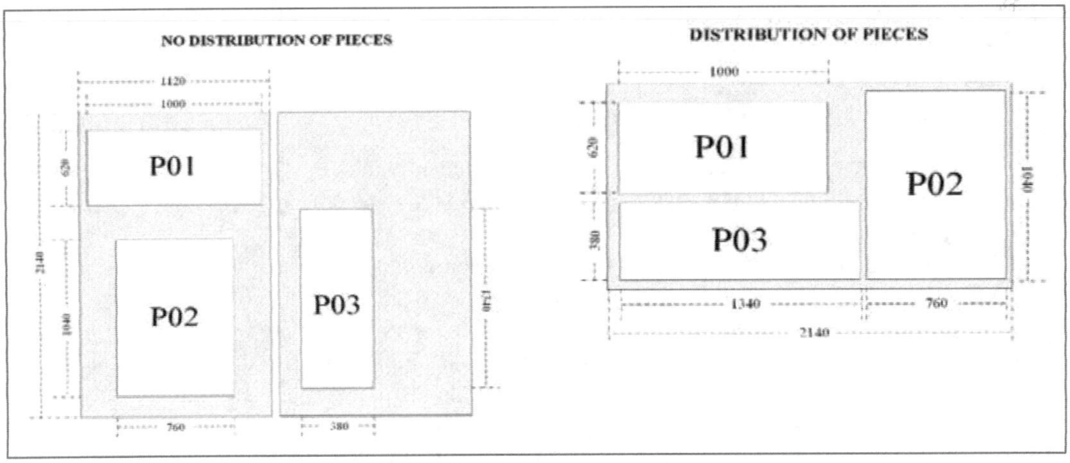

Figure 3. Distribution of pieces

Preventive Maintenance: The company had not defined a maintenance plan for the machinery used in the production process, so it resorted to using corrective actions when the failures occurred. Due to this, training sessions were carried out on simple maintenance actions with the purpose of allowing operators to perform them. In addition, the frequency of the activities to be carried out was defined, as well as what actions to be taken and the persons responsible. This was reflected in the increase in the average time between failure from 5 to 10 days due to the constant cleaning of the equipment, the placement of oil, grease and lubricants in the machines [15].

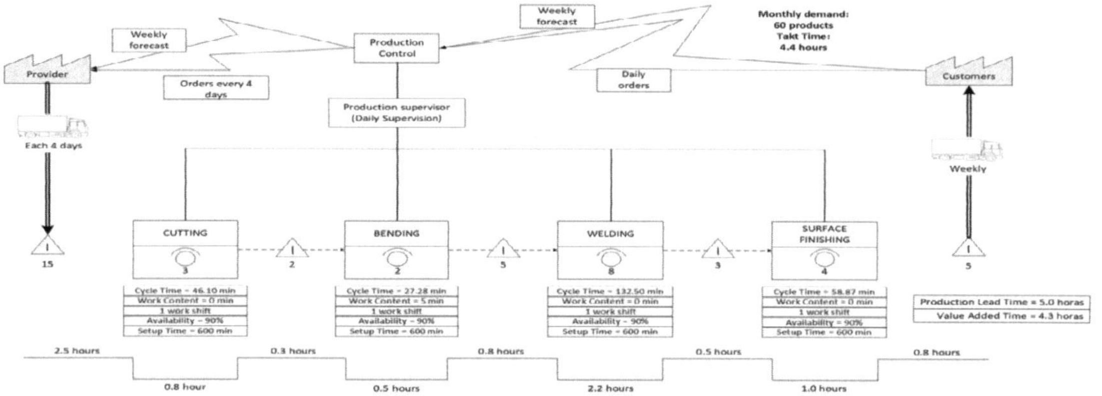

Figure 4. Proposed Value Stream Mapping

A value stream mapping of the future state of the production line was created taking into consideration the results obtained after the implementation of the Lean Manufacturing tools. In the new VSM, there is evidence of a reduction in the number of WIPs among the priority areas due to the implementation of work instructions that regulate activities in operations, reducing unnecessary movements and activities, seeking to increase productivity [5]. Likewise, the joint work of the 5S methodology with the standardization of the work achieved to reduce the cycle time of the operations and consequently the total manufacturing time from 5.8 to 4.3 hours.

5. Discussion

The results obtained after the implementation were the expected, starting with 5'S where were introduced instructional manuals and process management eliminating non value activities, this tool is jointed directly with standard work where manufacturing costs were reduced in 13% reducing wastes in cutting and bending processes. As well, preventive maintenance helped to focused on previous activities before machine failures and not after eliminating costs per machine stops and corrective maintenance.

6. Conclusion

This paper was focused in the synergy between different lean manufacturing tools and a soft one. At this way, this proposal generated a faster understanding by the employees including implementation support and giving an extra motivation.

On the other hand, in order to reach the goals, set, the materials that were not used in the fabrication of the products were eliminated. The standardization of the work through standardized procedures in the productive processes resulted in a 6% reduction in the waste index generated when cutting the stainless-steel plates. The preventive maintenance was also implemented reducing the incidences of faults in the cutting and bending machine, increasing the frequency from 5 to 10 days. These implementations formed the basis for a later application in the rest of the areas of the organization.

In general, the implementation of these tools demonstrated their relevance for achieving the proposed results generating a reduction of 13% of manufacturing costs.

References

[1] Thomé, A. M. T., Sousa, R. S., & Do Carmo, L. F. R. R. S. (2014). The impact of sales and operations planning practices on manufacturing operational performance. International Journal of Production Research, 52(7), 2108–2121

[2] Bellido, Y., Rosa, A.L., Torres, C., Quispe, G., Raymundo, C., "Waste optimization model based on Lean Manufacturing to increase productivity in micro- and small-medium enterprises of the textile sector", CICIC 2018 - Octava Conferencia Iberoamericana de Complejidad, Informatica y Cibernetica, Memorias, 1, pp. 148-153.

[3] Lacerda, A. P., Xambre, A. R., & Alvelos, H. M. (2016). Applying Value Stream Mapping to eliminate waste: A case study of an original equipment manufacturer for the automotive industry. International Journal of Production Research, 54(6), 1708–1720.

[4] Lastra, F., Meneses, N., Altamirano, E., Raymundo, C., Moguerza, J.M., "Production management model based on lean manufacturing for cost reduction in the timber sector in Peru", Advances in Intelligent Systems and Computing, 971, pp. 467-476

[5] Zhou, B. (2016). Lean principles, practices, and impacts: a study on small and medium-sized enterprises (SMEs). Annals of Operations Research, 241(1–2), 457–474.

[6] Andrade, Y., Cardenas, L., Viacava, G., Raymundo, C., Dominguez, F., "Lean manufacturing model for the reduction of production times and reduction of the returns of defective items in textile industry", Advances in Intelligent Systems and Computing, 954, pp. 387-398.

[7] Suárez - Barraza, M. F., & Ramis - Pujol, J. (2012). An exploratory study of 5S: a multiple case study of multinational organizations in Mexico. Asian Journal on Quality, 13(1), 77 - 99.

[8] Rose, A. N. M., Md. Deros, B., & Ab. Rahman, M. N. (2013). A study on lean manufacturing implementation in Malaysian automotive component industry. International Journal of Automotive and Mechanical Engineering, 8(1), 1467–1476.

[9] Mane, A. M., & Jayadeva, C. T. (2015). 5S implementation in Indian SME: a case study. International Journal of Process Management and Benchmarking, 5(4), 483.

[10] Guillen, K., Umasi, K., Quispe, G., Raymundo, C., "LEAN model for optimizing plastic bag production in small and medium sized companies in the plastics sector", (2018), International Journal of Engineering Research and Technology, 11(11), pp. 1713-1734.

[11] Khan, S. A., Kaviani, M. A., J. Galli, B., & Ishtiaq, P. (2018). Application of continuous improvement techniques to improve organization performance: A case study. International Journal of Lean Six Sigma.

[12] Azizi, A., & Manoharan, T. a/p. (2015). Designing a Future Value Stream Mapping to Reduce Lead Time Using SMED-A Case Study. Procedia Manufacturing, 2, 153–158.

[13] Gupta, S., & Jain, S. K. (2014). The 5S and kaizen concept for overall improvement of the organisation: a case study. International Journal of Lean Enterprise Research, 1(1), 22.

[14] Filip, F. C., & Marascu-Klein, V. (2015). The 5S lean method as a tool of industrial management performances. In IOP Conference Series: Materials Science and Engineering (Vol. 95).

[15] Antoniolli, I., Guariente, P., Pereira, T., Ferreira, L. P., & Silva, F. J. G. (2017). Standardization and optimization of an automotive components production line. Procedia Manufacturing, 13, 1120–1127.

[16] Sundar, R., Balaji, A. N., & Satheesh Kumar, R. M. (2014). A review on lean manufacturing implementation techniques. In Procedia Engineering (Vol. 97, pp. 1875–1885).

Creating a High and Homogenous Resolution Workspace for SCARA Based 3D Printers

Essence McClinton[1], Brianna Wylie[2], Dr. Carl A. Moore Jr.[3]

[1] Spelman College, Department of Computer Engineering, 350 Spelman Ln, Atlanta, GA, 30314

[2] Florida A&M University, Department of Mechanical Engineering, 1601 S Martin Luther King Jr Blvd, Tallahassee, FL 32307

[3] Florida A&M - Florida State College of Engineering, Department of Mechanical Engineering, 2525 Pottsdamer St., B373A, Tallahassee, FL 32310

Abstract. 3D printing is an additive manufacturing process that has an untold number of applications in fields as diverse as medical prostheses to military vehicle manufacturing not to mention its long term use in component prototyping [1][2]. With the creation of robotic arm based printers, the 3D printing process can be improved in terms of flexibility and time-efficiency, but with the potential trade-off being lower resolution in some areas of printer workspace. To counteract the resolution reduction, we are studying the use of continuously variable transmissions (CVTs) coupled to the robot's traditional revolute joints [3]. This paper will show that CVTs allow our SCARA-based robotic printer called DEXTER to achieve a resolution as good or better than a traditional desktop style 3D printer.

Keywords: Selective Compliance Assembly Robot Arms (SCARA), Spherical Continuously Variable Transmissions (SCVTs), 3D Printer Resolution, Dexter, MATLAB.

1. Introduction

1.1 Gantry 3D Printers

3D printing is an additive manufacturing (AM) process that builds customized parts layer-by-layer based on a computer-aided design file. The most common 3D printer configuration is a gantry-style or "H-frame" system. These printers use a set of orthogonal linear bearings to precisely position the extruder along the x and y axes of a print bed that may be raised or lowered. One benefit of a gantry style printer is that the extruder's workspace has a homogeneous resolution; meaning the smallest movements the extruder can make is consistent and independent of its position in the x-y plane. This characteristic results in a print resolution that can be maintained throughout the part. Building a homogeneous resolution printer with two or more independent extruders would enable the efficient printing of parts that contain unique material combinations including conductive materials within the layers [4]; however, H-frame and other parallel linkage configuration printers preclude the use of collaborating extruders (or other tools) without collision.

1.2 SCARA 3D Printers

Dexter, our prototype ambidextrous, multipurpose machine, has two SCARA (Selective Compliance Assembly Robot Arms) manipulators as shown in Figure 1 [5]. Each arm is made up of two links connected by two revolute joints. The base of the upper link is able to move vertically giving Dexter's extruders 3-degrees-of-freedom.

Figure 1. Dual SCARA 3D Printer "Dexter"

Dexter'sindependentroboticarmsprovidethedexteritynecessaryforcollaborativeandcollision free 3D printing and promise a host of benefits to traditional AM. For example, if each robotic arm is equipped with an extruder a part can be printed in nearly half the time. Or each extruder could print a different material, thereby facilitating the simultaneously integration of components that require unique substrates. Finally, a second arm could be equipped with a traditional tool permitting subtractive operations (e.g.: milling and sanding) to be done in tandem with the additive build process [6]. In short, Dexter's multi-SCARA configuration has the ability to greatly improve the capability and speed of AM.

The bases of Dexter's arms are located such that the arms' shared reach or workspace is maximized, and the motion of each set of arm links is controlled to avoid collision [7]. Like other 3D printers, Dexter's joints are controlled by stepper motors, an electric motor that moves in increments, or steps. Popular desktop style 3D printers use stepper motors with 200 steps per revolution. The stepper motor controller can increase that base number of steps through a process called microstepping resulting in extruder resolutions of 11 microns (0.0004 inches) like that of the MakerBot Replicator 2 [8]. However, the combination of links and stepper motors result in Dexter having a lower print resolution than traditional 3D printers, especially when one of its arm is outstretched. Consider that Dexter's stepper motors rotate its joints which in turn change the position of the links. To determine resulting extruder resolution ΔX_{res} the maximum steeper motor or joint esolution $\Delta\Theta_{res}$ must be multiplied by the manipulator's Jacobian J. The Jacobian matrix relates changes in joint velocities to tip of the arm velocities in Cartesian coordinates,

$$\Delta \mathbf{X}_{res} = \mathbf{J}\left(\Theta\right)\Delta\Theta_{res} \tag{1}$$

InsomelocationstheelementsoftheJacobianbecomequitelarge,especiallyneartheboundaries of the workspace, such that the smallest extruder motion is many times that of other locations. This change in positional accuracy is true for all articulated manipulators but is exacerbated by the use of stepper motors which have lower positional accuracy than the servo motor systems used in most manipulator systems.

1.3 Spherical CVTs

To improve Dexter's resolution, we are studying replacing its stepper motor joints with a spherical continuously variable transmission, or SCVT. The SCVT is a nonholonomic transmission based on rolling contact [9]. As shown in Figure 2, a SCVT consists of a sphere caged by four rollers: two drive rollers on top of the sphere and two steering rollers underneath and inside a bevel gear set for this particular SCVT.

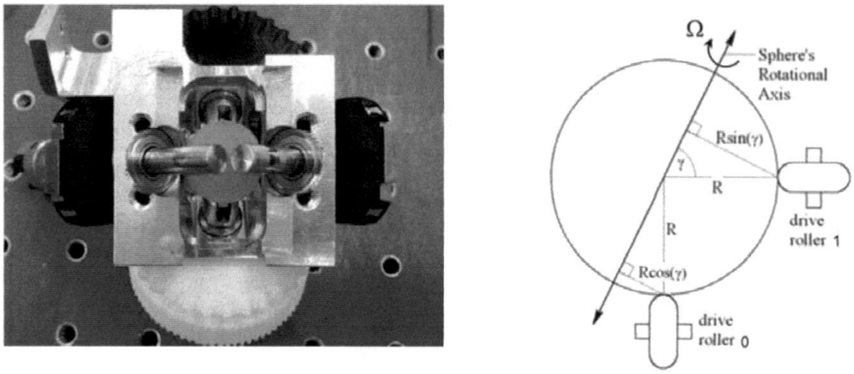

Figure 2. Left: SCVT with 0.5 in. sphere, drive roller shafts and steering roller gears. Right: Plane of the SCVT rotational axis which contains sphere's center and the contact points between sphere and drive rollers.

The sphere transmits angular velocity from one drive roller to the other at a ratio $\omega_1 : \omega_2$ determined by the sphere's transmission angle γ. The transmission angle is determined by adjusting the orientation θ of the steering rollers according to the following equation,

$$\frac{\omega_1}{\omega_0} = \tan(\gamma) = \frac{\sqrt{2} + \tan(\theta)}{\sqrt{2} - \tan(\theta)} \tag{2}$$

The SCVT transmission ratio is continuously variable up to a maximum goverened by friction. In practice we have achieved transmission ratios of nearly 6:1.

Using two SCVTs, our goal is to couple drive roller 1 from each SCVT to one each of Dexter's two joints. The second drive roller (drive roller 0) from each SCVT will be coupled to one another and driven by a stepper motor. Through computer control of both SCVT steering roller angles we can maintain any desired angular velocity ratio between Dexter's joints, thus permitting any desired extruder motion direction. The SCVTs provide for joint motion infinitely smaller than that of Dexter's original joint motors resulting in smoother nozzle paths and high resolution while still using a stepper motor input common to most 3D printers.

The objective of the present work is to show how SCVTs can be used in place of standard robot joints to increase the resolution of Dexter. This work will also investigate how resolution is affected through the increase of stepper motor steps for a traditional SCARA manipulator.

2. Experimental Methods

To investigate the effect of SCVTs on the resolution of a SCARA manipulator,MATLAB simulations were utilized in order to plot and compare the resolution graphs of Dexter using standard steeper motor joints to the resolution graphs of Dexter using SCVTs. In addition, multiple traditional SCARA simulations were run showing the effect of an increasing stepper motor steps per revolution.

Generally, to 3D print a part layer software is used to transform a CAD file into g-code which tells the printer nozzle where to move, how fast and at what temperature to heat the filament. We used MATLAB to read the g-code for a 'Calicat' part, a standardized file used for 3D printer calibration, and plot what would result if the output was sent to an actual 3D printer. The 'Calicat' plot in Figure 3 was simulated using approximately a 100x100 mm task space in MATLAB with a feed rate of 100 mm per minute. The program evaluated each line of the g-code, and plotted them as corresponding line segments in a Cartesian coordinate system which could be considered the build plate of a 3D printer.

IOP Conf. Series: Materials Science and Engineering **689** (2019) 012021 doi:10.1088/1757-899X/689/1/012021

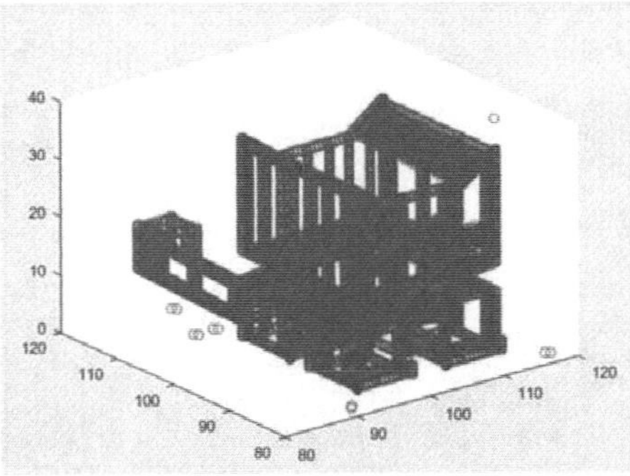

Figure 3. Calicat Plot

To visualize the effect of different resolution values, a single layer of Calicat was graphed in MATLAB assuming the nozzle was at the end of a 2-link SCARA-based 3D printer. Each link had a length of 203.2 mm and the base of the first link was attached to the origin of a x-y workspace. The robot's two joints were equipped with a 200 steps per revolution stepper motor. As the simulated nozzle followed the motion dictated by the g-code, the SCARA's inverse kinematics were used to calculate the change in joint angles for a given change in Cartesian nozzle position. These angle changes were subsequently interpolated to conform to the smallest possible joint angle change according to the available stepper motor steps.

The program graphs ellipses along the segment boundaries of a desired layer of the Calicat. The major and minor axes of each ellipse represents the resolution or minimum x and y-axis motions at that location for our example gantry style printer with a 11 micron resolution. A gantry printer has a homogeneous resolution so the major and minor axes of the resolution ellipses are everywhere equal forming circles. Finally the program graphs axes lines representing the resolution of the SCARA printer at the center of each circle graphed in the prior step. If the resolution of the SCARA printer is the same as the gantry printer then the lines will be contained within the circles printed previously. If the resolution is worse the lines will proceed outside of the circles and may not be parallel to the x and y-axis. These program steps were repeated three times for an increasing number of microstpes.

3 Results and Discussion

WesimulatedprintingtheresolutionellipseswithaSCARAbasedmanipulatorwitheithertraditional stepper motor joints or with SCVT equipped joints. For the traditional stepper motor cases we assumed a varying number of microsteps for the stepper motor. The first test was done with 16x microstepping resulting in a minimum joint rotation of 0.1125 degrees. Three additional simulations were done for 24x, 32x, and 64x microsteps resulting in minimum joint rotations of 0.075 degrees, 0.056 degrees and 0.028 degrees respectively. Next we performed a simulation with the SCVT equipped SCARA.

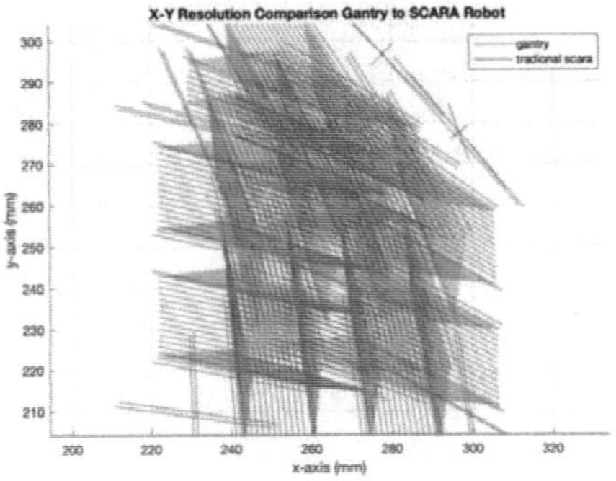

(a) 3200 Steps per Revolution

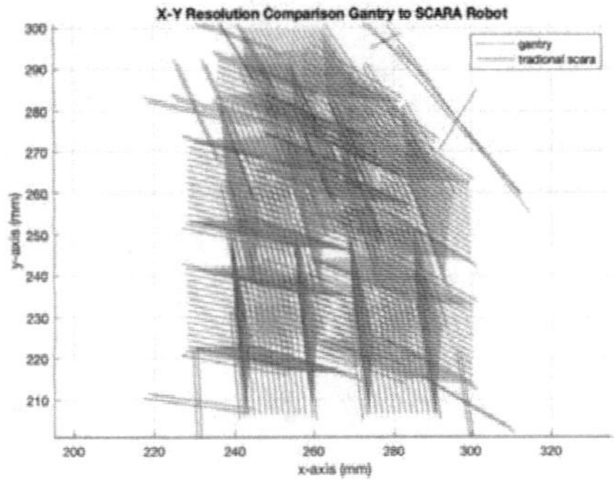

(c) 6400 Steps per Revolution

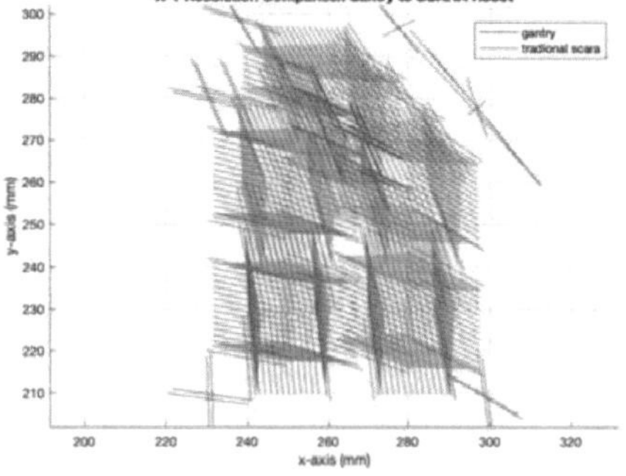

(b) 4800 Steps per Revolution

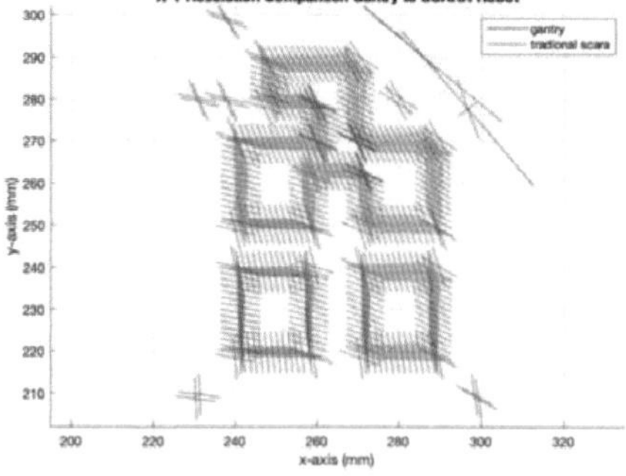

(d) 12800 Steps per RevolutionFig 4: Traditional SCARA Resolutions

Figure 4. Traditional SCARA Resolutions

3.1 Traditional SCARA Robot

According to the plots in Figure 4, a SCARA based 3D printer with standard stepper motors will result in much worse prints than is possible with a gantry-style printer like the Makerbot Replicator 2. The CalicatlayerresolutiongraphsimulatedusingaSCARArobotwith0.1125degreesminimum joint rotation (3200 steps per revolution) is shown in Figure 4a. The lines representing the major and minor resolution axes are very much longer and outside of the gantry resolution circles and not at the correct orientation. A SCARA with low resolution joints cannot produce perfectly horizontal or vertical lines at every location within its workspace. As the microstepping increases to 64x in Figure 4d, the ability to print improves drastically. There remains orientation error, especially at the distances furthest from the origin. As mentioned during the discussion of Equation 1, at these locations a SCARA based printer has non-homogenous resolution meaning that the smallest possible axis motions are dissimilar and along axes that may not be orthogonal.

As the microstepping increases, the lines on each plot became more contained, meaning that the resolution is improving and becoming closer to that of the gantry style 3D printer. However, it is important to note that higher microstepping is itself a control issue, often resulting in increased resolution but decreased accuracy.

3.2 SCVT-SCARA Robot

Upon using SCVTs at each of the SCARA's joints, the resolution increased dramatically. The size of the resolution axes in Figure 5 shows a many times improvement than even the best traditional SCARA plot that was shown above. The resolution also proved to be homogeneous in the x and y directions. Fromtheplot,itcanbeobservedthatnearlyalloftheprintlinesfromtheSCVTequipped SCARA printer were contained within the circles that represent the homogeneous resolution of the example gantry printer at each location, with the exception of the anomaly in the far-right corner which would be the farthest extension of the SCARA arm where nozzle motion is not possible in an arbitrary direction.

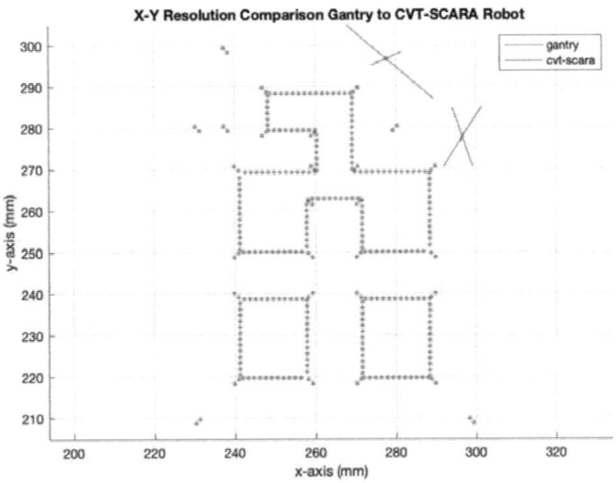

Figure 5. SCVT-SCARA Robot

4. Conclusion

The use of nonholonomic SCVTs improved the simulated print resolution of Dexter to many times that of a SCARA robot with traditional joint motors. The resolution also increases in homogeneity. Thus, usingSCVTscreatesthepossibilityofaSCARAbased3Dprinterwithgantryprinterextruder resolution while still using a stepper motor input.

For future work, we anticipate designing a SCARA based 3D printer with SCVT equipped joints. We are also interested in determining if a SCVT equipped 3D printer will excel at printing smooth curvature line segments.

Acknowledgments

I would like to acknowledge the RISE and MASS REU programs for funding this research project, as well as the National Science Foundation. I would also like to thank the adminstrators of the FSU AME Building for the use of the lab. A special thank you to Dr. Carl A. Moore Jr. and Brianna Wylie for their assistance on this research project.

References

[1] Smith P, Dombrowski M, 2017 Designing games to help train children to use Prosthetic Arms IEEE 5th Int. Conf. on Serious Games and Applications for Health (SeGAH) (Perth, WA) pp 1-4
[2] Suits, D 2019 3D printing technology enhancing logistics for Army Army News Service Feb 21 2019
[3] Moore C, Peshkin M, Colgate E 1999 A Three Revolute Cobot Using CVTs in Parallel Proc. ASME Int. Mechanical Engineering Congress and Exposition Nashville, TN
[4] Blackburn B, Camaratta M, Wachsman E 2009 Advances in Rapid Prototyping for Solid State Ionics ECS Transactions 16 pp 367-379
[5] Frketic J, Psulkowski S, Sharp A, Dickens T 2017 Procedia Manufacturing 10 pp 1087-96
[6] Luo R and Tseng P Carving 2D image onto 3D curved surface using hybrid additive and subtractive 3D printing process 2017 Int. Conf. on Advanced Robotics and Intelligent Systems (ARIS) (Taipei) pp 40-45
[7] McQueen K, Darensbourg S, Moore C, Dickens T, Allen C Efficient path planning of sec-ondary additive manufacturing operations 2018 Proc. 5th Int. Conf. Mechanical, Materials and Manufacturing (Orlando)
[8] MakerBot Replicator2 brochure, makerbot.com, 2019
[9] Brokowski, M, et al. Toward improved CVTs: theoretical and experimental results (2002) ASME Int. Mechanical Engineering Congress Exposition. (New Orleans)

IOP Conf. Series: Materials Science and Engineering **689** (2019) 012022 doi:10.1088/1757-899X/689/1/012022

Artificial Intelligence in the Aviation Manufacturing Process for Complex Assemblies and Components

McAndrew Ian, Vishnevskaya Elena, and Johnson Michael

[1] Capitol Technology University, Chichester, England
[2] Embry Riddle Aeronautical University, Bitburg, Germany
[3] The Boeing Company, North Charleston, SC, USA

Email: irmcandrew@captechu.edu; navarrj1@erau.edu; Michael.E.Johnson17@Boeing.com

Abstract. Aviation manufacturing is at the leading edge of technology with materials, designs and processes where automation is not only integral; but complex systems require more advanced systems to produce and verify processes. Critical Infrastructure theory is now used to protect systems and equipment from external software infections and cybersecurity techniques add an extra layer of protection. In this research, it is argued that Artificial Intelligence can reduce these risks and allow complex processes to be less exposed to the threat of external problems, internal errors or mistakes in operation.

1. Introduction

Commercial aircraft are now complex systems with parts, sections and components designed and produced globally. Commercial aircraft Assembly is more than classic assembly of bolting or joining parts together, systems are integrated, and automation is a foundation of all stages in manufacture. Many machine tool companies supply 'turn-key' systems that are controlled remotely to produce sub-assemblies. Directly, these are operated on-site, and local maintenance personnel ensure any concerns addressed immediately. When larger scale problems need sorting the software fixes may be completed remotely; this is potentially an infection point for a problem. Fig 1. below, shows a nosecone sub-assembly for an aircraft and indicates the complexity of the manufacturing both in terms of size and process.

Figure 1. Automation used in assembly for an aircraft nosecone.

2. Problems for consideration

There are several principal problems facing complex aviation automation that might result in damage, problems or long-term reliability concerns. On the basic level it could be a simple scratch on a component surface that leads to a fatigue crack of one where tooling is broken that results in risks to human lives. These can be categorized as: disgruntled employee, errors in operational use, deliberate external or internal control to cause immediate damage, deliberate external or internal control to cause damage at a later date and combinations of the latter two [1]. This research paper and research focuses on the last three stated. In theory, the production could be halted or even sabotaged by hackers; to damage the equipment. Such actions will make the equipment unusable or change tolerances and inspection values, which will approve components that need rejecting.

Aviation manufactures are aware of these concerns and have not been slow to react. Cybersecurity is key and applied at all stages of the development; both internally and externally. Likewise, Critical Infrastructure planning is now fundamental for all new equipment and software. Collectively these both attempt to prevent access and, if gained, access it will limit what could be changed [2]. They are not foolproof and as experience shows not a guarantee. Artificial intelligence offers possibilities to add a third and more effective layer of protection to the complete process. Additionally, neural networks are systems where decision making can be achieved to localize error detection or problem identification. System that can and should be used to compliment the detection are available but seldom, if ever, fully applied. This paper discusses the advantages and how they can work. It is a seminal research with the aim of incorporating manufacturing quality into the decision making process.

3. Review of techniques used to protect process

Cybersecurity has evolved in the last decade to be something that was considered and added with passwords to a whole industry that is adapting to ever more advanced attacked from governments and individuals. As expectations increase the use of passwords have been shown to be ineffective and many still use passwords that are predictable. Reports have shown that systems could be immediately be made more robust by 80% is Admin access was prevented. This research is more on the utilization than practical solutions. Systems needs minimum characters, special characters and numbers but still these are 'cracked' and by ever more advanced methods by hackers that could be from foreign governments, multi-nationals or individuals. Wi-Fi hopping is just one example of how *hackers* try to keep ahead of protections systems. Even government defense systems have been breached at times and these are with budgets for Cybersecurity far in excess of most industrial organizations. The reported security breaches are not publicized as much and probably not a main target currently; however, this is likely to change. Many manufacturers could find that their complete system is externally *infected* and unable to operate [3]. Regardless of the size of the company a Cyber-attack could potentially bankrupt any company or hold it to ransom for a release fee. As Cyber protection increases and the acceptance of prevention is not high, the concept of Critical Infrastructure (CI) has been added to systems. The concept is that if access gained to a system by unauthorized personnel then the scope of the damage possible is minimized and the system will recognize these actions and shut down their access or isolate to with the intention to stop. These plans can be physical, for example, access to areas and for remote access that a stealth approach is adopted to hide systems and limit what they can actually control and change [4]. CI is adding protection to manufacturing that compliments Cybersecurity, collectively, it creates two barriers; these are still effective but not making a fool proofing to give total confidence. Most literature will address defense, banking, food, transport and utilities but not commercial manufacturing. As a sector, manufacturing is not as advanced as others are and leaves vulnerabilities, see Fig 2 below.

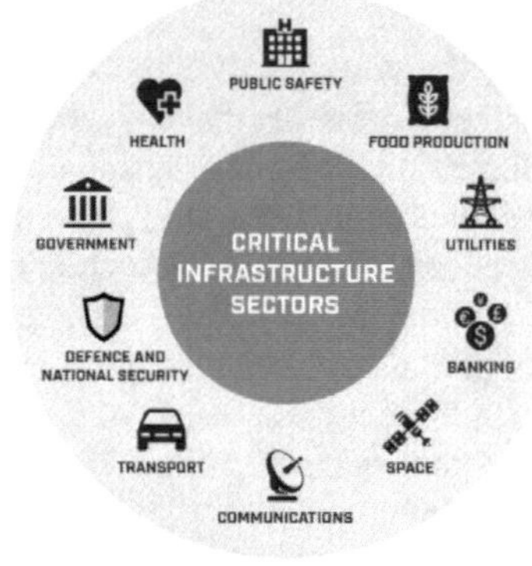

Figure 2. Principal Critical Infrastructure sectors.

Parallel to Cyber protection and Critical Infrastructure the subject of Artificial Intelligence (AI) is being researched and advanced for control systems in many sectors and uses. Manufacturing needs systems that not only make the system work but in ways that will enhance the operation and efficiency. Neural networks process information in the same biological way as a human brain. The interconnections allow for advantages and speed of processing [5]. Artificial neural network is one where there are many simple units of nodes each may have their own local processor. All this local units are connected by unidirectional channels, connections, which carry numeric data. All units work on a local level and with that data from these local unidirectional connections. Learning by example is the approach used as in the biological learning of humans. For example, if sending a signal to stop a motor operating is not effective, the signal will be sent by other connections to achieve the desired output.

Training is the basis for artificial neural networks and is generally classified as supervised or unsupervised. The former is where input and output values are given, and these are compared to make a logical decision from the local data [6]. When no inputs or outputs given it is unsupervised. These systems are autonomous as work logically as a child might identify an object. Unsupervised is ideal where human use skill to recognize but are not always able to clarify the decisions made each time [7]. Patterns are identified and remembered that the next time the correct decision is made instantly. Manufacturing is an example of where operators learn the process and identify errors or omissions. A good example is when we walk away from a car and close the door without looking back; the sound or lack of sound tells us of the output closed correctly.

—Neural Networks, NN, are massive parallel computation structures. This allows calculations to be performed at a high speed, making real-time implementations feasible, an important aspect of both protecting against cyber-attacks and cyber attackers attempting penetrate systems. Artificial neural networks, ANN, are useful in control systems where conventional approaches are not satisfactory. In fact, artificial neural systems are widely used nowadays in a variety of industrial control systems [8]. A large number of applications of ANN to robotics, failure detection systems, nonlinear systems identification and control of nonlinear dynamical systems have been proposed in recent years. Neural network models are known as an efficient and accurate tool for simulating manufacturing processes and as manufacturing systems become more complex their need is greater [9].

Neural networks is one of the applications in the cognitive science of AI. Neural networks are computing systems that are modelled after the brain's mesh-like network of interconnected processing elements (neurons). Interconnected processors operate in parallel and interact with each other. See figure 3 below.

These systems:

- allows the network to learn from the data it processes;
- recognizes patterns and relationships in data; and
- assists in the effective operation of any system.

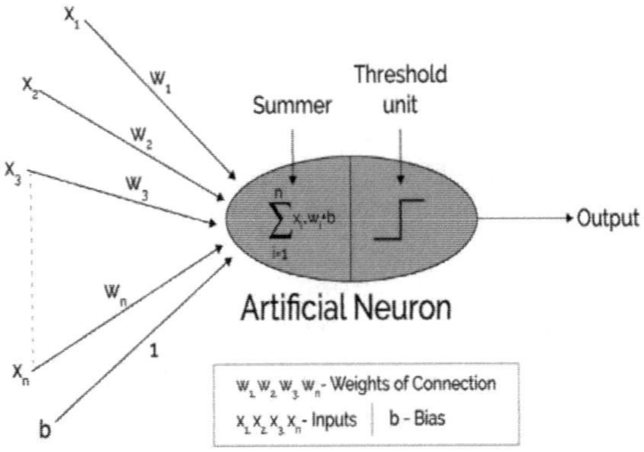

Figure 3. How Artifical Neural Networks operate.

In the 1980s, a new development in programming paradigms appeared that we know now as artificial neural systems, ANS. This was possible with faster computer processing times and that the microelectronics could incorporate localized IC to have input and output devices remotely in a system to suit the needs and designs [10]. It is only recently that these are becoming more widespread in all walks of industry. They work by:

- being based on the fundamental way the brain processes its information from its senses and knowledge stored;
- models solutions by training simulated neurons connected in a network to make the output dependable;
- similar to an analog computer using simple processing elements connected in a highly parallel manner;
- processing elements perform Boolean / arithmetic functions in the inputs; and
- key feature is associating weights with each element.

Initially, they were used in military and advances applications due to the costs and technical support. Nowadays, through the developments and drive for power systems they are becoming more widespread. It has been suggested that these will become core of all systems when security and intelligence needed [11]. In aviation manufacturing the final assembly is usually with main components made globally. Boeing have supplies in south America, Australia, Europe and these components can be major parts. For example, AIRBUS, has its wings flown in from Wales and assembled in France and German factories. Quality at all stages needs controlling and what is critical is that variations between parts and within parts needs to be consistent. Monitoring this and ensuring that they are protected from direct or indirect data corruption is now paramount. Inspection is not quality, using the full range of technology is now the minimum and that does not guarantee success and robustness from the problems stated above. Advantages of ANS:

- Nets can extrapolate and interpolate from their stored information;
- Nets have plasticity; and
- Excellent when functionality is needed long-term without repair in hostile environment – low maintenance.

ANS are not suited for number crunching or problems requiring optimum solution as is often needed in manufacturing, for example, direct measurements and comparisons to technical drawings. They do not make inferences but searches for underlying patterns [12]. Artificial neural networks is a popular methodology with lots of practical and industrial applications, from prediction models and control of turbines to industrial ANN control of calcinations processes. Identifying patterns can be a statistical science on its own leading to the subject now addressed as data analytics.

The goal is not to build a human brain or even try to fully replicate, rather to support analysis of complex processes for modern systems, manufacturing and applications.

4. Analysis of artificial neural networks

Manufacturing in the civil aviation sector has been behind the advances of the principal CI industries as shown in fig. 2. The potential for crippling problems in the short, medium and long term is significant and most companies have accomplished minimal research. An external hacker that gained control or access to a manufacturing system could program the robots and machines to make movements at speed, which result in the destruction of the equipment and prevent any further work. Research has shown and argued Artificial Neural Network overseeing the system (ANN), Cyber and CI aspects that are embedded from the start in the design may offer benefits but fail to identify fully [13]. The AI aspects need to be incorporate at the launch stage of production to record, analyze and store this data that constantly can be used to compare and make decisions and at all stages of the lifespan the aircraft is produced. First, unsupervised training will be needed if used in addition for the start of production; although supervised training might be needed and useful for improving the system faster. Any security infections have the potential to control the process, alter tolerances or accept rejected parts. How this integrates in individual systems cannot be determined until a review related to Cyber and CI is applied to a specific process and forms the basis for the future work.

In Fig. 4 below shows an advanced automation by a robot that is controlled externally, locally by the system and also maintenance when needed. If controlled externally and the cyber protection breached, and critical infrastructure compromised there are several main problems or risks that need evaluation. Typical manufacturing systems could be changed that results in tolerances widened and thus rejects made and accepted [14]. If these are allowed to be used on an aircraft the potential for a disaster is compounded. Risk might evaluate these situations; identifying these risks if random or controlled is near impossible. Today's governments, companies and individuals are threats to aviation manufacturing. This ignores, as stated above, disgruntled and former employees.

Considering current manufacturing system architecture where Cyber and CI are integrated to control a complete system; the result is one where specified task, protocols and stages are completed in order and manner. If an artificial neural network is added as a top-level control that may initially be set to be a supervised system and eventually an unsupervised system that its role will be multi-faceted [15]. First, supervised ANN will be used to identify the basic concerns and activities to meet requirements. This would include a closed-loop on inspection to individual inputs for unsupervised teaching. Secondly, patterns, changes in patterns and outliers in the process can be identified for possible concerns and corrections. Thirdly, the times when software updates, remotely or locally, can be recognized and local concurrence at a high authority level will ensure any breaches are minimal or identified immediately. Finally, there should be no accessible externally to the ANN to prevent cyber-attacks. This latter comment is easy to state but forms the pivot of the problem to be solved.

Perhaps, its greatest role will be in unsupervised training for trends and comparison in component manufacture, assembly and testing. A full in-depth investigation is currently not complete and this research reports on the feasibility and implications of how the architecture can be employed that meets all three needs.

Figure 4. Automation as a sub-assembly within a larger system controlled locally and remotely.

Fig 5. below shows another aviation example where two identical processes are used. Consider the comparison possible for this data with unsupervised training of the information. Not just in manufacturing to pass inspection but long term. The life, maintenance, reliability and breakages of these aviation parts will be recorded for 30 years whilst commercially used. Given this data can be cross-referenced for its complete life variations in dimensions, interactions of measurements and success offers immediate, medium-term, long-term and life feedback of the parts to truly enable quality to be integrated at every level of use and predict future problems based of numerical analysis not probability.

Figure 5. Identical manufacturing processed.

Manufacturing in commercial aviation is a process that is expensive, extremely long-term in planning and one where mistakes and errors have huge consequences for all [16]. Every time a new aircraft is planned the finances needed are values approaching above the complete turn-over financially for several years and profits need forecasting over 30 years lifespans. Boeing's B 737 Max may yet result in significant restraints to the company, if this error was externally caused a company might never recover financially. Unlike other industries the aviation success depends significantly on reputation for safe products. The complexity now is at a level where control must extend beyond classic ways and explore advanced ways where the tasks can offer more than human control and dynamic feedback to support and identify needed actions. The potential to be inclusive in the design and quality stages can further enhance the complete system. Systems need evolving more to adapt and improve as technology changes and advances. Civil aviation manufacturing is such a sector and without adopting new technology a company will become globally redundant.

5. Summary

In this paper it has be argued that Critical Infrastructure and Cybersecurity techniques offer protection to aviation manufacturing up to a certain level. Artificial Intelligence with the focus on artificial neural network offers a level that adds protection and a unique dimension to learn and prevent both deliberate external and internal actions and identify trends to improve the process. Furthermore, this approach has benefits as feedback to suppliers, utilizing lessons from start-up problems to assist quality and reliability. As risks to systems increases then systems must be prepared and ever improving. Artificial Neural Networks is one dimension that can assist and offer additional benefits.

References

[1] Acemoğlu, D. and P. Restrepo (2017) Robots and jobs: Evidence from the US. https://voxeu.org/article/robots-and-jobs-evidence-us (Accessed on 30 July 2019)
[2] Cybersecurity Tech Accord (2018). Available from: www.cybertechaccord.org (Accessed 29th June, 2019).
[3] O'Brien, J & Marakas, G, (2006) Management Information System, McGraw Hill ISBN 0-07-111629-X
[4] Nydegger, R., & Enides, C. (2017). The psychology of work: Changes in the 21st century. International Business & Economics Research Journal, 3rd Qtr. 16(3), 197-203. Retrieved July 9, 2019, from DOI https://doi.org/10.19030/iber.v16i3.9993
[5] Haykin, S. (1994). Neural Networks, a Comprehensive Foundation. Macmillan, New York, NY.
[6] Beale, R. and Jackson, T. (1990). Neural Computing, an Introduction. Adam Hilger, IOP Publishing Ltd : Bristol. (ISBN 0-85274-262-2)
[7] Dayhoff, J. E. (1990). Neural Network Architectures: An Introduction. Van Nostrand Reinhold: New York.
[8] S. Jauhar, B. Chen, W. Temple, X. Don, Z. Kalbarczyk, W. Sanders, D. Nicol. "Model-based cybersecurity assessment with nescor smart grid failure scenarios". In 2015 IEEE 21st Pacific Rim International Symposium on Dependable Computing (PRDC) 2015 Nov 18 (pp. 319-324). IEEE.
[9] Buschle, M., Holm, H., Sommestad, T., Ekstedt, M. and Shahzad, K., 2011, June. A Tool for automatic Enterprise Architecture modeling. In International Conference on Advanced Information Systems Engineering (pp. 1-15). Springer, Berlin, Heidelberg.
[10] S. Noel, J. Ludwig, P. Jain, D. Johnson, R. Thomas, J. McFarland, & B. Tello. "Analyzing mission impacts of cyber actions (AMICA)." In NATO IST-128 Workshop on Cyber Attack Detection, Forensics and Attribution for Assessment of Mission Impact, Istanbul, Turkey, 2015.
[11] Lemaire, L., Lapon, J., De Decker, B. and Naessens, V., 2014. A SysML extension for security analysis of industrial control systems. In 2nd International Symposium for ICS & SCADA Cyber Security Research 2014 (pp. 1-9). BCS Learning & Development Ltd.; North Star House, North Star Avenue, Swindon, SN2 1FA, UK
[12] Wang, Y. N., Lin, Z. Y., Liang, X., Xu, W. Y., Yang, Q., & Yan, G. F. On modeling of electrical cyber-physical systems considering cyber security. Frontiers of Information Technology & Electronic Engineering, 17(5), 465-478, 2016.
[13] D. Liu. "Systems engineering: design principles and models." Boca Raton, FL: CRC Press 2015
[14] North Carolina State University Laboratory for Analytic Science website: https://ncsu-las.org/research-areas/sensemaking/ (Accessed on 30 July 2019)
[15] Le, T. (2018) Bridging Cybersecurity Skills Gap Through AI. Available from:https://www.scmagazine.com/home/opinions/bridging-the-cybersecurity-skills-gap-through-ai/
[16] Kagermann, H., Helbig, J., Hellinger, A., and Wahlster, W., "Recommendations for Implementing the Strategic Initiative Industrie 4.0: Securing the Future of German Manufacturing Industry; Final Report of the Industrie 4.0 Working Group," Forschungsunion, 2013. medium, provided the original work is properly cited (CC BY 4.0).

Multi-scale model retrieval for case recommendation in manufacturability optimization

Zhi Li[1], **Danchen Zhou**[1] **and Bo Zhang**[1]
[1] Institute of Machinery Manufacturing Technology, China Academy of Engineering Physics, China

Email: 15681163908@163.com

Abstract. This paper proposed a multi-scale retrieval framework to help optimize manufacturability. The case based workflow integrates both geometry and rules with model-based definition. At conceptual stage, defect models are categorized and global retrieval is initialized to narrow down cases as defects references. At detail stage, local feature retrieval finds similar defect feature with pre-defined rules, to adapt to manufacturability optimization. Intersection feature detection and decomposition are explored to improve generality.

1. Introduction

Similar to simulation that optimizes structural performance, manufacturability analysis could optimize manufacturing feasibility. Optimized model would reduce cost and repetition of manufacturing. Thus such tool is essential to industry. While there are lots of engineering failures cases, managing these resources is valuable to industry companies. Similar cases could be references to decide how failures portion on the target model should be fixed. Given that medium of engineering design upgrade to model based definition (MBD)[1], which is composed of geometric and non-geometric information, like product and manufacturing information (PMI), existing retrieval method should adapt to the case recommendation task. To retrieve proper reference portions, recommender systems need to find out critical face geometry, and to compare relationship and/or the properties of local elements, which is different from existing search system or feature recognition system. Given that there might be intersection features in practice, handling intersection features is also needed for generality.

2. Related Work

If design is modelled by valid machining features, it is of good manufacturability and is fit for principle of design for manufacturing (DFM). Tools like DFMpro[2] were proposed to reduce workload of manual check. Machining feature library is predefined like hole etc., and text based semantic rules are coded to recognized features. Similar to DFMpro, Siemens NX Checkmate[3] aims at standard validation, including formatting and company standards, and general best practices. Definition of rules is also text-based not link to models. Configuration space(C-space) modeling[4] approach validates space relationship of setup between product and tools. Wang et al.[5] proposed slice of cavity for cutting tool selection, but it cannot adapt to other manufacturability analysis task. Fox et al. proposed split of light weight model and combine manufacturability of each part to a degree, heat map was introduced to visualize how different regions affect cost[6], which is apparent to users.

Case based reasoning(CBR) is to solve new problems by remembering previous similar situation and by using information and knowledge of that situation. For example, Jiang et al.[7] proposed a fuzzy similarity-based rough set method for case-based reasoning and explore its application in cutting tool selection. Guo et al. [8] implemented a CBR system for injection mold design based on ontology. Hashemi et al.[9] proposed a case-based reasoning approach to design machining fixture.

Since retrieval is the first step of CBR, similarity assessment is essential. There are many approaches in content based similarity assessment, such as histogram based[10], feature based[11], graph based[12] etc.. In this paper, a multi-scale retrieval approach was proposed to solve the MBD case recommendation problem for manufacturing optimization.

3. Overview

Framework is illustrated in figure 1. Given a new design model, global retrieval identifies which class it belongs to where repository was indexed in advance. Then, defects are located in the specific class by local feature retrieval. Optimization rule stored on referenced MBD model are returned, parsed to driven target model update, to optimize manufacturability. If the updated model is of reusing value, it could be stored in the repository to enlarge the capacity of the framework. In short, it contains conceptual and detail stage and follows the CBR principle, related to model reuse and knowledge fusion. This paper focuses on the locating stage, not interactive definition and adaptation of rules.

For every defect, it could be geometry defect, non-geometry defect or combination, since the MBD model may contain geometry model and PMI information (figure 2). These defects are based on geometry faces because solid model spaces are defined by face boundaries (B-rep), and PMI are attached to faces. The geometry basis could be a single face, multiple faces, or discrete faces with some relationship. For example, a drilling simple hole is of ill manufacturability if the length–diameter ratio is larger than a threshold, the geometry basis is a single face. An intersection feature like milling non-rectangle slot, however, is composed of multiple faces, where the manufacturability requirement is related to all of the faces and the depth value of slot should not be too large to be accessible. If the turning corner radius is smaller than bottom corner radius, it is of ill manufacturability, where the geometry basis are discrete cylindrical faces. Therefore, precise face locating is required.

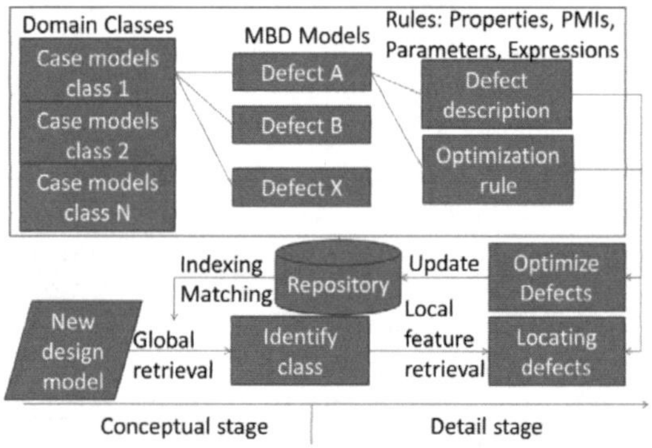

Figure 1: Overview of the framework.

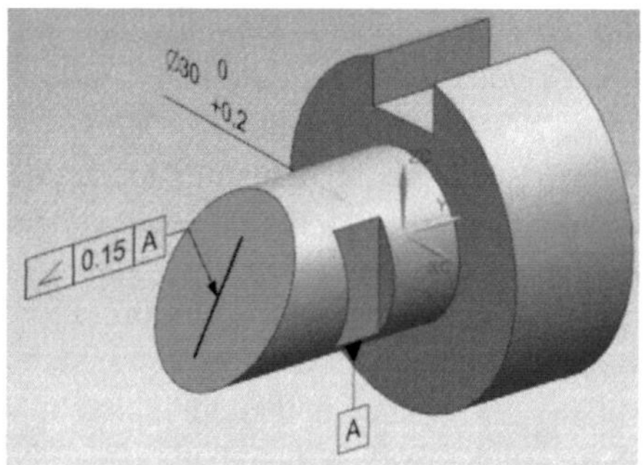

Figure 2: Model based definition.

4. Multi-Scale Solid Model Retrieval

4.1. Initial Filtering and Detailed Retrieval

Assume that in the same class similar parts could share more probability of similar defect feature, a classified defect parts repository could automate the class identification of the target model by evaluating MBD similarity. The process is as follows:

1. Collect typical models, mark class information of each model, to initialize the model repository.
2. Set similarity assessment algorithm to be the distance function of classifier.
3. Adopt k-nearest neighbour (KNN) algorithm to mark the target model.

Since MBD model is mainly composed of solid model (B-rep) and PMI, the global similarity should be composed of these two kinds of information. We extend hierarchical partition graph (HPG) [13] to measure MBD similarity, which conform to rotational, transformation and rotational invariant.

Main idea of HPG is to partition a model into several parts. Each partition is a maximum set of faces with coherent concave/convex connections, which can be organized in a hierarchy. If two models are similar, the similarity will distributed on description in level of detail. For example (figure 3), the hierarchy is composed of a tree of resolution level (see dashed lines) and some adjacent graph (see solid lines). While the tree preserves global topology consistency by 'parent-children' relationship between each resolution level, the adjacent graph preserves local topology consistency by 'sibling' relationship between each partition node. Connections of part node are called 'seed wire'. Detail of the seed wire recognition, partitioning, assembling algorithms, and matching process please refer to [13].

Since non-geometry information is also defined on faces, it can be stored as face property in the detail similarity calculation stage. For example, distance between the numerical value-based property like dimensions, tolerance etc. (which should be of the same PMI type) can be measured by normalizing the Euclidean distance, which is absolute ratio between target-reference value difference, and maximum-minimum difference with in the PMI type. Engineering notes are treated as structured XML[14] for comparison. Since the MBD model are defined from the design platform like Siemens NX, where PMI are stored as XML, the values and sub-XML can be extracted and parsed using NX Open API. After considering non-geometry similarity, formula 2 in [13] could be extended as follows:

$$S_{leaf} = \sum_j \omega_{leaf}^{in} \times (\omega_g \times \mathrm{D2_{sim}} + \omega_{ng} \times XML_{sim}) \quad (1)$$

Where $\mathrm{S_{leaf}}$ represents non-geometry similarity influence leaf partition of HPG, $\omega_{\mathrm{leaf}}^{\mathrm{in}}$ is the ratio between matched faces' area of $\mathrm{j_{th}}$ node and total matched faces' area of all nodes in this resolution. $\mathrm{XML_{sim}}$ is non-geometry similarity where PMI defined on all faces of each partition. $\omega_{\mathrm{g}}$and ω_{ng} are weights for geometry and non-geometry similarity, respectively. It requires $\omega_{\mathrm{g}} + \omega_{\mathrm{ng}} = 1$.

In the same model class, locating relevant case to check manufacturability is to find relevant faces. We explored face properties matching to tackle this problem[15]. The main idea is increase matching accuracy by extending highly detailed local symmetry transition of around faces. Since B-rep can be represented as winged-edge data structure[16], every edge has two 'wing' faces, each of which has two 'wing' edges. Convexity change of these two edges, is used to define symmetry of the wing face:

1. Convex symmetry : both 'wing' edges are convex
2. Concave symmetry: both 'wing' edges are concave
3. Non-symmetry: convexities of two 'wing' edges are different.

As illustrated in figure 4, there are two evaluating edges. The edge convexity is convex (left) and concave (right). Each edge is intersected by two highlighted adjacent faces ('wing' faces). For each wing face, two connected edges ('wing' edges) sharing the same loop with the evaluating edge. Because convexities of both wing edges are convex, symmetry of face related to the evaluating edge can be defined as convex symmetry. And the symmetry across evaluating edge is unchanged.

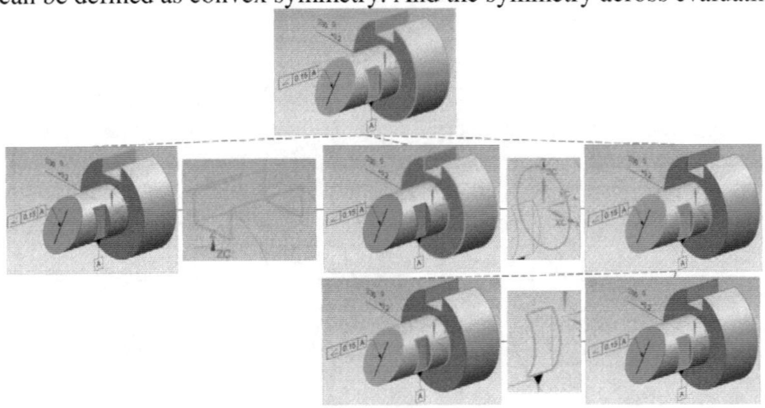

Figure 3: Hierarchical Partition Graph.

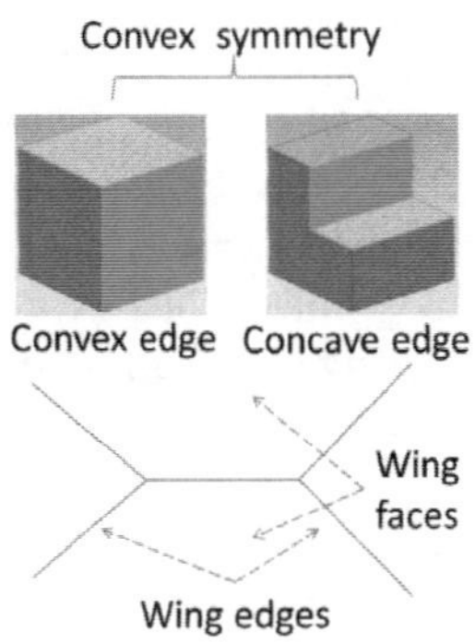

Figure 4: Edge symmetry

While convexity can be defined on every edge, so as the symmetry to all the faces, therefore the matching criteria of targeting face could be strictly set to be matched both the face property, its boundary edge properties and even the intersection point properties. The basic properties for matching can be parsed in advance. Such as the face normal, centre point, solid face type (plane, cylindrical, spherical etc.), edge orientation (forward or reverse), solid edge type (liner, circular etc.). In addition, the relational property can be extended by the edge convexity and face symmetry. The matching rules:

1. Rules for faces: Basic solid face type and extended property like face symmetry.
2. If face rules meets, further check whether its boundary edges rules meet to edge properties.
3. Edge rules: solid edge type, edge convexity and extended edge symmetry changes (next section).

4.2. Handling Intersection Features

We explored geometric reasoning based intersection detection and decomposition for reusing existing defect cases on simple feature to optimize intersection features' manufacturability [15]. Based on the observations that features are around concave edges, we further define edge symmetry change to distinguish intersection feature hint, which is defined as concave edges whose symmetry changes along its 'wing' faces. As illustrated in figure 5, there are four concave edges. The bottom two edge's adjacent faces' symmetry are unchanged, the top two edges' are changed, therefore the latter two edges are intersection feature hints, which can be used for intersection feature decomposition.

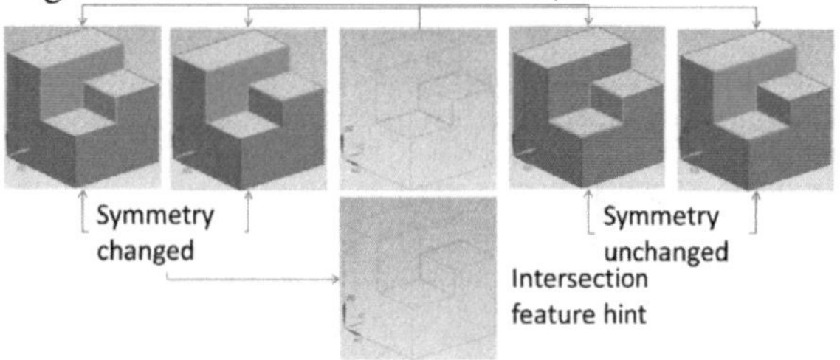

Figure 5. Intersection feature hint.

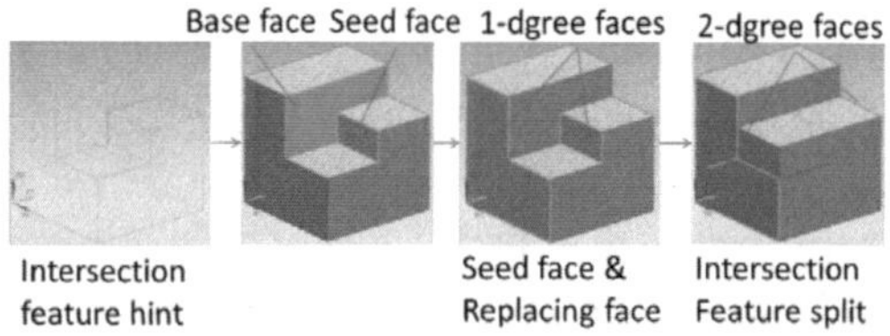

Figure 6. Feature decomposition.

In figure 6, selecting one splitting edge from intersection feature hint, the larger face of the two highlighted 'wing' edge is base face(defined as a face that has at least two splitting edges), and the smaller one is seed face. 1-degree faces are defined as one-step expanding faces of seed face, and adjacent to the base face. 2-degree faces are one-stop expanding faces of 1-degree faces adjacent to base face. Replacing face is selected from 2-degree faces. For multiple 2-degree faces, the selection criteria is maximizing the dot product value between seed face normal and a vector, which isfrom seed face centre to replacing face centre. The 'replace face' operation would preserve 1-degree adjacent relationship between seed face and replacing face, which can be implemented with NX direct modelling API. Updated model will reduce complexity of intersection feature hint. Iteratively perform the 'replace face' until the rest concave edges are not splitting edges. The intersection feature could be divided into multiple single features, which could enlarge the reusability of existing repository.

5. Experimental setup and initial results

While defining manufacturability defects and optimization rules are based on Siemens NX RDDV module in .prt format, repository for identifying class type is composed of 101 engineering MBD files classified in 11 categories, like flange, socket, brick etc. every class contains multiple typical defects.

Initial filtering contains offline indexing and online matching stage. Since the extended properties like symmetry and symmetry change are also related to the face properties, they are initialized as exploring of the B-rep topology. As illustrated in figure 7, after initial filtering, the target model get its class type of bracket, and the further searching of defects are in a most relevant defect cases.

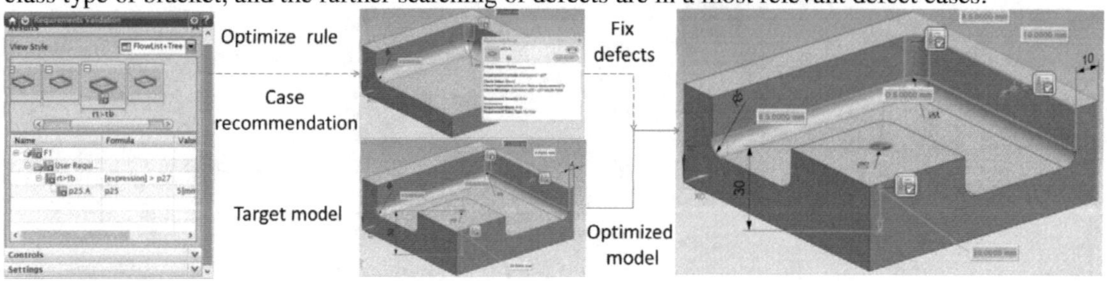

Figure 7: Manufacturability case.

After detailed retrieval, highlighted defect areas are found to be with relation to three recommended cases, where through hole is related to one face, thickness of wall is related to two faces, and chamfer is related to multiple discrete faces. Taken chamfer as an example, optimizing rule is 'the radius of turning corner should be larger than radius of bottom corner', which is defined on the defect case with NX RDDV module. Similarly, while length of hole should be less than 5 times of diameter in best practice, wall thickness should be larger than 5 mm. After retrieve requirement rule defined on related faces, target models' face parameters are optimized after approve of the recommendations.

6. Conclusion

In this work, a multi-stage case recommendation approach combining initial filtering and detailed retrieval are proposed. Geometric reasoning is used to implement the proposed method. A case based manufacturing framework is supported by MBD model retrieval, and implementation demonstrated its effectiveness. Based on the preliminary study, we would like to further explore intersection feature recommendation, to help automate the complex manufacturability optimization.

Acknowledgments
This work is supported by national science foundation of China (NSFC), the grant No. is 51705485.

REFERENCES

[1] Alemanni M., Destefanis F., and Vezzetti E., 2011, "Model-Based Definition Design in the Product Lifecycle Management Scenario," International Journal of Advanced Manufacturing Technology, 52(1–4), pp. 1–14.
[2] "DFMPro"[Online].Available:https://dfmpro.geometricglobal.com/. [Accessed: 28-Jan-2019].
[3] Anderson T., [Online]Using Teamcenter Requirements to Validate NX Designs.
[4] Nelaturi S., 2012, "Configuration Modeling," PhD Thesis, University of Wisconsin-Madison.
[5] Wang Y., Ma H., Gao C., Xu H., and Zhou X., 2005, "A Computer Aided Tool Selection System for 3D Die/Mould-Cavity Nc Machining Using Both a Heuristic and Analytical Approach," International Journal of Computer Integrated Manufacturing, 18(8), pp. 686–701.
[6] Fox K. A., Massoni B., and Campbell M. I., 2018, "A Method for Creating a Manufacturing Cost Heatmap Based Purely on Part Geometry," (51760), p. V02BT03A030.
[7] Jiang Y., Chen J., and Ruan X., 2006, "Fuzzy Similarity-Based Rough Set Method for Case-Based Reasoning and Its Application in Tool Selection," International Journal of Machine Tools and Manufacture, 46(2), pp. 107–113.
[8] Guo Y., Hu J., and Peng Y., 2012, "A CBR System for Injection Mould Design Based on Ontology: A Case Study," Computer-aided design, 44(6), pp. 496–508.
[9] Hashemi H., Shaharoun A. M., and Sudin I., 2014, "A Case-Based Reasoning Approach for Design of Machining Fixture," The International Journal of Advanced Manufacturing Technology, pp. 1–12.
[10] Mehrdad V., and Ebrahimnezhad H., 2015, "3D Object Retrieval Based on Histogram of Local Orientation Using One-Shot Score Support Vector Machine," Frontiers of Computer Science, 9(6), pp. 990–1005.
[11] Kwon S., Mun D., Kim B. C., and Han S., 2017, "Feature Shape Complexity: A New Criterion for the Simplification of Feature-Based 3D CAD Models," International Journal of Advanced Manufacturing Technology, 88(5–8), pp. 1–13.
[12] Tao S., Wang S., and Chen A., 2017, "3D CAD Solid Model Retrieval Based on Region Segmentation," Multimedia Tools & Applications, 76(1), pp. 103–121.
[13] Li Z., Zhou X., and Liu W., 2015, "A Geometric Reasoning Approach to Hierarchical Representation for B-Rep Model Retrieval," Computer-Aided Design, 62(0), pp. 190–202.
[14] Tekli J.,Chbeir R., and Yetongnon K., 2009, "An Overview on XML Similarity: Background, Current Trends and Future Directions," Computer Science Review, 3(3), pp. 151–173.
[15] Li Z., Lin J., Zhou D., Zhang B., and Li H., 2018, "Local Symmetry Based Hint Extraction of B-Rep Model for Machining Feature Recognition," (51791), p. V004T05A009.
[16] Baumgart B. G., 1972, Winged Edge Polyhedron Representation, No. STAN-CS-320, STANFORD UNIV CA DEPT OF COMPUTER SCIENCE.

AUTHOR INDEX